To Interpret the Earth

Ten ways to be wrong

To Interpret the Earth

Ten ways to be wrong

Stanley A. Schumm

Department of Earth Resources, Colorado State University
and
Water Engineering and Technology, Inc.

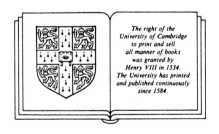

The right of the
University of Cambridge
to print and sell
all manner of books
was granted by
Henry VIII in 1534.
The University has printed
and published continuously
since 1584.

CAMBRIDGE UNIVERSITY PRESS

Cambridge

New York   Port Chester   Melbourne   Sydney

Published by the Press Syndicate of the University of Cambridge
The Pitt Building, Trumpington Street, Cambridge CB2 1RP
40 West 20th Street, New York, NY 10011, USA
10 Stamford Road, Oakleigh, Melbourne 3166, Australia

First published 1991

Printed in Great Britain at the University Press, Cambridge

British Library cataloguing in publication data
Schumm, Stanley A. (Stanley Alfred) *1927–*
A scientific approach to earth science.
1. Earth sciences. Methodology
I. Title
550.1

Library of Congress cataloguing in publication data
Schumm, Stanley Alfred
A scientific approach to earth science: ten ways to be wrong /
S. A. Schumm.
    p.   cm.
Includes index.
ISBN 0 521 39507 0
1. Earth sciences – Methodology.
QE40.S38 1991
550'.28–dc20   90–37675   CIP

ISBN 0 521 39507 0 hardback

Contents

Contents

Preface

The basis for this monograph was an invited lecture at the University of Washington in 1982 on the validity of the use of analogs in geologic interpretation. In subsequent years the topic was expanded through a series of lectures presented to both undergraduate and graduate students in geology and watershed science at Colorado State University. One purpose of the lectures was to attempt to describe a 'scientific method' or an orderly procedure by which the students could proceed with their assignments and their research goals.

The preparation of these lectures was surprisingly difficult and time consuming, because of the diversity of opinion concerning this topic and the surprising lack of concern by earth scientists. Therefore, it seemed worthwhile to put certain observations and considerations concerning the scientific method in a format that could be of use to young earth and environmental scientists and students in the organization of their research. This book is, therefore, an attempt to clarify confusion regarding the scientific method and to stress an approach rather than a method of investigation. In addition, a major part of this effort is to describe ten specific problems faced by the researcher, as one works to explain the complexity of this planet. It differs from the other books of this type because it not only attempts to discuss the scientific method appropriate to earth science, but it also considers in some detail specific research problems that make it necessary to think less of a method and more of an approach.

Finally, some general solutions to the problems are proposed. It is assumed that being aware of these problems will be of assistance to those involved with complex natural systems. This awareness may not aid in the solution of the problems, but it will instill caution and an appreciation of the pitfalls facing one who is engaged in research.

Although discussion of the scientific method is based largely on the classic geologic publications of Gilbert and Chamberlin, the term earth science is frequently used instead of the word geology because the problems are of more general concern, and they can be more readily visualized when the examples are of surficial features and processes. Therefore, what is presented, is relevant to many types of environmental studies.

Finally, this short work should not be considered as a foray into the philosophy of science. It should only be read by young earth and environmental scientists and students and not by philosophers who would undoubtedly be aggravated by its lack of depth. It is, indeed, a very personal account of the scientific approach and the problems associated with it in the earth sciences.

Acknowledgements

I confess that in spite of 20 years of support by National Science Foundation, and US Army Research Office, this small monograph was written without their direct support. Nevertheless, much of what is written is based upon research carried out with the generous assistance of these agencies during my tenure at Colorado State University. Significant typing and drafting assistance was provided by Water Engineering and Technology, Inc. The typing expertise of Saundra Powell is greatly appreciated, as was her ability to read my writing.

Stella Leopold is largely responsible for my involvement in these topics because of her invitation to discuss the validity of analogs in earth science.

The participants in the Colorado State University Geomorphology Seminar (Spring 1988) were forced to review an early version of the manuscript and the criticisms and suggestions for improvement from graduate students John Pitlick, David Jorgensen, Benjamin Hayes, Daniel Levish, Chris Williams, Robert Rogers, Elliott Lips, and Deborah Anthony were of considerable help during revision of the manuscript. In particular John Pitlick and David Jorgensen provided detailed reviews of the manuscript, and they contributed many useful comments. I also thank my colleagues for their suggestions for improvement of the manuscript as follows: Frank Ethridge, Colorado State University, Guillermina Garzon, University of Madrid, Z. B. Begin, Geological Survey of Israel. I am particularly grateful to David Sugden, University of Aberdeen, for his review and especially for his suggestion that Popper's and Chamberlin's ideas could be reconciled and that the ten problems could be grouped into three categories.

1 To diagnose the Earth

It is not an easy thing to say what the scientists are doing or what science is.
Feibleman, 1972, p. 1

Diagnosis is considered to be a medical term, meaning to distinguish or identify a
disease, but as used in the title the word has broader implications. Diagnosis is 'an
analysis of the nature of something' (W. Morris, 1981). In medical practice a diagnosis involves taking the history of a complaint, examination of the patient,
identification of the problem, statement of the diagnosis and finally the statement
of a prognosis. Therefore, the physician describes the condition, predicts the
course of the disease, and prescribes a cure. In each case, the physician is dealing
with a single individual, and he is trying to apply generalizations to one person.
This is very similar to the practice of the earth scientist, who in most instances is
dealing with a singular situation, perhaps a single but complex feature at or near
the earth's surface such as a mountain range, an outcrop, a river or a hillslope.

The procedure followed by the physician during diagnosis is a scientific
method, and, in fact, the objectives of this short work are to discuss briefly the
scientific method in earth science and then to stress the problems encountered in
its application. However, the first thing that a reviewer of the literature on the
scientific method discovers is that there is little agreement about it, and the
opinions are extreme. Sarton is emphatic that 'The great intellectual division of
mankind is not along geographical or racial lines, but between those who understand and practice the experimental method and those who do not understand and
do not practice it' (Mackay and Ebison, 1977, p. 134). Bertrand Russell (1961,
p. 243) concludes that 'whatever knowledge is attainable, must be obtained by
scientific methods, and what science cannot discover, mankind cannot know'.
Conversely, Medawar (1979b) a Nobel Laureate, says, in effect, that there is no
scientific method. He concludes that if an effective scientific method existed then
scientists would be more successful in their endeavors than they are, and if a
method existed that leads a scientist with certainty to the truth then there is no
excuse for not solving scientific problems. This perhaps facetious conclusion can
be coupled with his statement that 'the scientific method is an enormous potentiation of common sense' (Medawar, 1979a, p. 27). Potentiation is defined as a
powerful accumulation. So, according to Medawar, the scientific method is
simply the use of common sense, but according to Boorstin (1983, p. 294; see also
Nagel, 1961, pp. 1–14) 'Nothing could be more obvious than that the earth is
stable and unmoving and that we are the center of the universe. Modern Western

1

science takes its beginning from the denial of this common sense axiom . . . Common sense, the foundation of every day life, could no longer serve for the governance of the world'. Perhaps by common sense Medawar means trained and critical thought.

In 1986, Sigma Xi (1986), the Scientific Research Society, sent a questionnaire to its membership. One statement related to science and the scientific method, as follows: 'The word science is often invoked as if it meant a particular "thing" comprised of scientists, public and private laboratories, publications, and government agencies. For me, however, "science" connotes a process or procedure for making inquiries about our world and for evaluating the hypotheses these inquiries generate'. Of the scientists who responded to this questionnaire, 95% agreed with the statement that science connotes a method for making inquiries about the world. This suggests that in the minds of most scientists it is the method employed in carrying out their research that distinguishes science from other human endeavors.

Scientists in many fields give a great deal of attention not only to the method by which their activities are carried out, but also to the success of these activities (Susser, 1973; Harvey, 1969; Sayer, 1984). Medicine provides a good example, where, of course, there is great public concern with success, and there are a number of books available on differential diagnosis, which describes the manner by which a physician proceeds to diagnose an illness and to prescribe for it.

NATURE OF EARTH SCIENCES

Most earth scientists do not find philosophical discussions of their field very interesting. In fact, many scientists treat the philosophy of science with 'exasperated contempt' (Medawar, 1984, p. 132). This is understandable because the terminology of the philosopher is as difficult to comprehend to the uninitiated as is the language of scientific specialists. Indeed, geologists seem to exhaust their philosophical proclivities in speculating about the nature of geology as a science (Simpson, 1963; Watson, 1966, 1969; Van Bemmelen, 1961; Bucher, 1936, 1941; Kitts, 1963b; Spieker, 1965), and they are less interested in discussing their role as scientists (Kitts, 1977, p. XI). With the exception of T. C. Chamberlin (1890, 1897), G. K. Gilbert (1886, 1896), Douglas Johnson (1933, 1940), and Hoover Mackin (1963) geologists tend to go about their scientific endeavors without giving much thought to the manner in which they proceed.

This may be the reason that Mackin (1963, p. 137) in a footnote to his excellent paper on rational and empirical methods in geology, complained that geology is scarcely mentioned in the large literature on the history and philosophy of science. Nevertheless, since then Kitts (1977) and Watson (1966, 1969) have written extensively on the subject, and Gilbert's work has attracted considerable attention (Yochelson, 1980; Pyne, 1978, 1980; Kitts, 1973) as has Chamberlin's (Laudan, 1980; Pyne, 1978) and the work of Hess (Franklin, 1980).

2

It is generally agreed that geology is a derivative science (Scriven, 1959; Pantin, 1968; Bucher, 1941) that utilizes information and concepts from astronomy, chemistry, physics, and biology (Fig. 1.1). If as David Kitts (1974) concludes, geologists are not much inclined to discuss their role as scientists, part of the problem is that they generally attempt to apply this diverse information to singular or particular events or things, and therefore, their explanations may often be weak in comparison to those of other sciences (Kitts, 1963b, p. 25; Nagel, 1961, pp. 547–606; Simpson, 1963). Also, it has been stated that because geology is historically oriented and because it deals with such great complexity, during long periods of time and over large areas, that it is different from other sciences. Usually this means that geology differs from chemistry and physics and that somehow the difference is demeaning to the geologist. This is a peculiar attitude that seems to have been

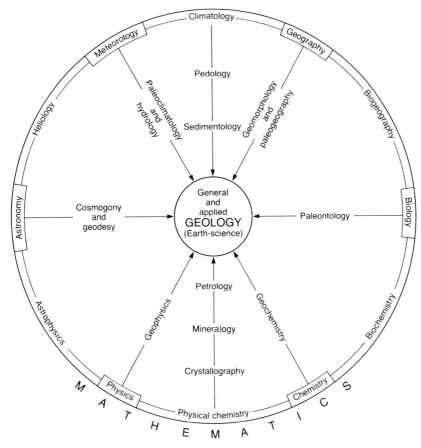

Figure 1.1. Geology as a derivative science (from Van Bemmelen, 1961).

3

generated by the inability of the geologist to produce quantitative 'laws of nature'. Indeed, the consideration of both vast spans of time and large areas is not common to other physical sciences (Simpson, 1963; Watson, 1969) with the obvious exceptions of astronomy and cosmology. The geologist, therefore, is involved in the study of 'complex natural experiments conducted on a large scale in both time and space' (McKelvey, 1963). The experiments with which geological history confronts us are neither reversible nor repeatable, and they are accomplished on a scale of time and space that precludes as a matter of course exact reproduction. Moreover, they cannot be directly observed; but they must be reconstructed historically (Bubnoff, 1963, p. 3). This is the fundamental distinction between geology and most other sciences. The earth scientist deals with complex systems that function over long periods of time, and each system, although not unique, may be singular (Nairn, 1965; Watson, 1969). That is, each system is different from similar systems, and each may reflect processes no longer active at a particular location (e.g. glaciation). Hence, the opportunity for reproducibility and falsification is minimal. Therefore, it is not surprising that geologic and geomorphic predictions may have 'low resolution' and may be weak in comparison with other sciences (Pitty, 1982).

The variability of geologic processes and rates, during long periods, and the limited number of situations that can be sampled, creates a problem not encountered when dealing with closed systems, isolated variables, experimental results, and the statistical analysis of very large numbers of measurements (Hagner, 1963, p. 235). Geology, as with biology, psychology, history, economics, geography and engineering is an 'irregular subject' in the sense of Scriven (1959) because there can be considerable error in the application of generalizations to specific cases (see also Kitts, 1963a; Leopold and Langbein, 1963; Grinnell, 1987, p. 27). But depending upon time and scale this is true of other sciences. For example, in meteorology it is possible to predict the general climate of a region based upon past records, but it is not easy in the short term to predict changes in the weather. In physics and chemistry, if the behaviour of the single atom is considered rather than the average behaviour of many, the ability to predict is also very low. The engineer uses factors of safety to compensate for the difficulty of specific predictions. Unfortunately, the need to predict in earth science is usually for specific cases (site stability, well locations).

Geology also contains within itself a great dichotomy between historically-oriented and process-oriented research. Bucher (1941) uses the terms 'timeless' and 'time-bound' to describe the differences. Timeless knowledge involves 'the search for general properties and patterns of behavior or "laws" that characterize the object and their reactions to each other. These properties and laws apply always, everywhere. They are independent of the stream of time'. Time-bound knowledge is concerned with a specific object (river system, hillslope, mountain range, sedimentary deposit) and its change with the passage of time.

Strahler (1954) prefers to distinguish between historical geology (time-bound knowledge) and physical or dynamic geology (timeless knowledge) on the basis of repeatability or the probability of recurrence of a particular state or form. He proposes that 'historical investigation be defined as referring to the analysis of complex states having very small probabilities of being repeated, that is, to states of low recoverability. Dynamic investigation in the same context refers to the analysis of states having a high degree of probability of being repeated, such analysis leading to the formulation of laws of general validity.' Most of us operate in both of these areas of earth science, some more than others, but this situation sometimes leads to problems of communication and interpretation (see Chapter 3).

The following statement by D. A. Pretorius (1973), although written for the economic geologist, is equally applicable to many aspects of geology and is an admirable statement of the problems facing the earth scientist attempting post-diction:

> It is the nature of the history of the earth that a geologist has available to him only partial information. Occasional lines from disconnected paragraphs in obscurantist chapters are what can be read. Violence in the handling of the book through time has caused many of these chapters to be ripped and reassembled out of context. That the gist of the early chapters can be deciphered at all is a credit to perseverance and imagination not always associated with other sciences. The geologist operates at all times in an environment characterized by a high degree of uncertainty and ornamented with end-products which are the outcomes of the interactions of many complex variables. He sees only the end, and has to induce the processes and the responses that filled the time since the beginning.

As an undergraduate, the definition of geology that I learned was that it is 'the science of the earth as revealed in rocks'. This definition of geology has been replaced by a somewhat more inclusive definition in the glossary of the American Geologic Institute (Gary *et al.*, 1972), as 'the study of the planet earth'. The definition continues as follows: 'It is concerned with the origin of the planet, the material and morphology of the earth, and its history and the processes that acted (and act) upon it to affect its historic and present forms.' This definition also implies that the main concern is the explanation of present conditions and the interpretation of history. However, later in this rather lengthy description, prediction is acknowledged, and it is stated that 'All of the knowledge obtained through the study of the planet is placed at the service of man, to discover useful materials within the Earth; to identify stable environments for the support of his constructed arts and utilities; and to provide him with a foreknowledge of dangers associated with the mobile forces of dynamic Earth, that may threaten his welfare or being.' The newer definition involves the geologist in extrapolation not only into the past but also into the future.

5

EXTRAPOLATION

In the area of applications some earth scientists are very concerned with ongoing processes of fluvial, aeolian, coastal and glacial erosion and deposition, ground-water movement, seismic activity, active tectonics, etc., and therefore, these geologists predict. Indeed, with the increasing importance of environmental geology, engineering geology and geomorphology, prediction has been empha-sized in the past few decades. Therefore, earth scientists must extrapolate from the present not only to the past, but also to the far future, especially when the stability of hazardous-material disposal sites is being evaluated.

Extrapolation, of course, involves the projection of known information or relationships to the unknown. This involves both extrapolation to the future, which is prediction, and extrapolation back in time, which is retrodiction or post-diction. Figure 1.2 is an attempt to show schematically how studies of modern processes and their effects provide a description and explanation of the present, which can be used by analogy to develop models of the past and future. Past changes and conditions in turn, influence the present, and, therefore, the future.

It is possible to learn a great deal about the present. Measurements of erosion and deposition can be made and the dynamics of aeolian, fluvial, glacial, coastal and marine systems can be investigated. Information on the morphology and dynamics of these systems permits the development of a model of contemporary situations. Extrapolation from the present to both the past and future is then possible (Fig. 1.2). However, it is obvious that if both the present and the past are known, then prediction is enhanced because the record is extended by historical information. Therefore, when present conditions are known and understood and when the history of the situation has been established, predictions can be made with some degree of confidence. This is especially true when the time spans that are involved are relatively short. For example, a history of river behavior during the last 100 years, when coupled with data on the present morphologic, hydro-

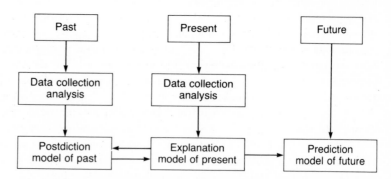

Figure 1.2. Diagram illustrating how information from the present and past is used to construct models of the past, present and future (from Schumm, 1985).

Table 1.1. *Time spans of concern to various earth science disciplines*

The near past represents historic time and the far past all of prehistoric time. The near future represents the next 50 years. The far future from 50 to 10 000 years.

Past		Present	Future	
Far	Near		Near	Far

```
◄───────── Geology ──────────────────────────────────────►
              ┌────────── Geography ───────►
              ├────────── Civil engineering ───────►
◄─ Paleontology ┼──────────►
◄─ Stratigraphy ┼──────────►
◄─ Sedimentology ──────────►
◄─ Economic Geology ──────────►
◄─ Petrology ┼──────────►
◄─ Mineralogy ┼──────────►
◄──────────────── Geophysics ──────────►
              Environmental geology ────────────────────────►
              Engineering geology ─────────────────────────►
◄─────────── Hydrogeology ───────────────────────────────►
◄────────── Geomorphology ───────────────────────────────►
```

logic and sedimentologic situation, can produce predictions for at least 50 years if conditions remain unchanged during that time.

It should be noted that the term prediction is used in two ways in science. The first is the standard definition, which is to foretell the future. The second is to develop a hypothesis that explains a phenomenon. For example, a petroleum geologist predicts the occurrence of gas and oil, but the prediction relates to the occurrence of something that was formed in the remote past. We can predict that a geologic structure (fold, fault) exists based upon outcrop patterns, but the structure may have formed millions of years ago. Therefore, in some instances the term prediction is used synonymously for postdiction or extrapolation. This common second use of the term prediction may be why the terms postdiction and retrodiction are infrequently seen in geologic literature. They are included in the geologist's use of the term prediction.

Table 1.1 is an attempt to illustrate the predictive time spans of concern for some of the major earth-sciences disciplines, and for some geologic specialties. Many are clearly historical, whereas others deal primarily with the present and the near future. The near future can be defined as the next 50 years, which is the civil engineer's requirement for the lifetime of many projects. Nevertheless, the geomorphologist and environmental geologist are not only concerned with the present and the past but also with the near and far future. They are called upon to estimate landform stability for the disposal of toxic and radioactive materials, and in some instances there have been requests to estimate stability of sites for 10 000 years. This is clearly impossible, and in any such attempt the worst-case scenario must be developed, but it is clear that prediction is now an important part of some aspects of geology. This is an important development because many of the definitions of science implicitly include the assumption of prediction. However, explanation of phenomena is as valid as prediction in science (Kitcher, 1982), and explanation itself may lead to prediction.

Sayer (1984) recognizes that prediction and explanation are valid goals of science, but he also identifies two other types, which fall between the two extremes of prediction and explanation. These he terms non-explanatory predictions and non-predictive explanations. For example, the search for oil is an exercise in non-predictive explanation. The petroleum geologist can explain all of the necessary conditions for the existence of oil but he cannot predict that it will be present. Drilling provides the proof. Furthermore, many techniques of prediction do not provide an explanation of the phenomena predicted. For example, we might be able to predict the afternoon weather by determining the number of people carrying raincoats or umbrellas to work, but no explanation of weather patterns would necessarily result. We may not understand river behavior, although the sand that is moving through the channel can be predicted as a function of the cube of flow velocity. Therefore description can be predictive or non-predictive.

The earth scientist's goal is to dispel ignorance about our planet. This has become increasingly important because, as Lewis Thomas (1979, p. 16) recognizes, human beings are swarming over the surface of the earth changing everything and meddling with it, 'making believe we are in charge, risking the survival of the entire magnificent creature'. He believes that we could be excused for this behavior on the grounds of ignorance, and indeed, 'in no other century of our brief existence have human beings learned so deeply, and so painfully the extent and depth of their ignorance about nature'. In fact, 'It is this sudden confrontation with the depth and scope of ignorance that represents the most significant contribution of twentieth-century science to the human intellect' (L. Thomas, 1979, p. 73). One can either be encouraged or discouraged by this remark. Why, with all our effort in the last century, have we not been successful in developing a thorough understanding of our planet?

The answer lies in the complexity of the system, which makes complete descriptions and accurate predictions difficult. As a result, other scientists and engineers who want a definitive statement about future landform changes or geologic stability are often disappointed by the vagueness of the earth scientist's predictions. The civil engineer, however, forgets that the factors of safety that are used to insure some degree of permanency of a structure reflect the difficulty of determining exactly the material properties and conditions for which the structure has been designed.

If geological predictions are perceived as being too general, is there a problem with the methods employed? The approach employed by earth scientists is no different from that employed by other scientists. However, one should remember that even Nobel Laureates have great difficulty in describing the scientific method, and in fact, some philosophers of science deny its existence (Feyerabend, 1975, 1978). Indeed, if there is not a strict method of science, part of the reason is because of the great variety of problems encountered in science and the different modes of attacking a specific problem.

DISCUSSION

In this chapter I have attempted to describe the difficulty faced by the earth scientist who is investigating complex open systems and attempting to explain them and to postdict and predict their behavior. Theirs is a noble effort made difficult by the multiple variables acting and the time spans involved. Others who are dependent upon these investigations for answers to practical problems may feel that the answers are not sufficient, but they must be made aware of the problems faced by the earth scientist, which are the problems that will be discussed in Chapter 3.

It seems to me that a discussion of these problems is of value to an understanding of the methods of our complex science. Therefore, a major objective of this work is to consider the problems inherent in extrapolation of modern conditions and relations to the past and to the future (postdiction and prediction). This is done with the expectation that such a discussion may help to explain why pronouncements about natural systems are sometimes vague, why extrapolation can carry a high probability of error, and why there is disagreement about the scientific method.

In fact, if one accepts the statement by Grinnell (1987, p. 27), a cell biologist, that 'Investigators know that reproducibility of experiments is a requirement of scientific research. While one can raise questions about unique events, only recurring events can be subjected to scientific investigation', then there is no earth science. Of course, the great understanding of Earth and time produced by geological investigation refutes any such suggestion, but Grinnell's comment emphasizes that there is great difficulty in applying a simple method to a system that has such variety and great complexity as does the planet Earth, and therefore, in Chapter 2 we will consider geologic methods in the light of this complexity.

9

2 Scientific method

It is true that much time and effort is devoted to training and equipping the scientist's mind, but little attention is paid to the techniques of making the best use of it.

Beveridge, 1957, p. iv

In order to attempt to resolve the obvious disagreement among scientists concerning the scientific method, some understanding of what is implied by the words is needed, and therefore, definitions of the words *science* and *method* are required. The word science is derived from the Latin words scientia, knowledge, and scientificus, making knowledge (Little *et al.*, 1964, p. 1806). Hence, science involves the business of discovery and the production of new knowledge. It is an activity that supposedly creates objective knowledge. That is, science is the attempt to learn the truth about those parts of nature that are explorable (Chargaff, 1978, p. 156).

The word method is defined in *The Oxford dictionary* (Little *et al.*, 1964, p. 1243) as 'a special form of procedure adopted in any branch of mental activity, whether for exposition or for investigation'. Hence, method is a way of doing anything according to a regular plan. So, if there is a method there is a procedure or a systematic way of pursuing a goal. Interestingly enough, the scientific method is not defined in this very comprehensive dictionary, but in *The American heritage dictionary* (Morris, 1981, p. 1163) it is defined as 'the totality of principles and processes regarded as characteristic of or necessary for scientific investigation, generally taken to include rules for concept formation, conduct of observations and experiments and validation of hypothesis by observation and experiments'. The three major components of this definition are, 1. concept formation, which involves generation of hypotheses, 2. observation and experiments, which involve the procedures of data collection, and 3. testing of hypotheses by observation and experimentation. This short statement adequately describes a scientific method. Data collection and testing are critical components of this definition, but because of the diverse nature of 'science', it is not possible to state a single method that applies to all sciences.

METHODS

Beveridge (1957) says that research is a complex and subtle task. It is difficult to teach and, therefore, one should just go do it. In effect, one should learn by one's mistakes in carrying out research, but nevertheless, he wrote his book *The art of scientific investigation* to assist the young scientist. The title of his book implies that there is a great deal of subjectivity in the activities of science, and indeed, if we take

10

the definition of method to mean a formal procedure or a specific way of proceeding then this subjectivity invalidates the definition of scientific method. Techniques and procedures are going to differ for the different sciences, and when one considers the great range of scientific endeavors it is no wonder that there is disagreement about whether or not there is a scientific method.

Nevertheless, we know that certain methods are not appropriate, such as the methods of Lysenko: 'distortion of facts, demagoguery, intimidation, dismissal, reliance on authorities, eyewash, misinformation, self-advertising, repression, obscurantism, slander, fabricated accusation, insulting name calling, and physical elimination of opponents . . . ' (Medvedev, 1969, pp. 191–2). In most scientific circles such methods would not be effective, but with Stalin's support they were effective in the USSR. Such techniques will not personally enhance a reputation, and they are not recommended for the pursuit of science.

Beveridge (1980, p. 55) does outline a scientific method in five steps, but he also includes some problems that are associated with each as follows:

1. recognition and formulation of the problem, but the statement of the problem may be incorrect,
2. collection of relevant data, but in many cases it is difficult to know what data are relevant,
3. induction of a hypothesis, but induction is unreliable,
4. deduction and testing of the hypothesis, but there can be error during testing,
5. revision of the hypothesis to give a more comprehensive explanation of the results, or starting again at step 3 by stating additional hypotheses, but the results may be misleading.

Earth scientists should disagree with Beveridge's outline, which involves development and testing of a single hypothesis, because the 'method of multiple working hypotheses', as stated by Chamberlin (1890, 1897) involves the formulation of as many hypotheses as possible, which are then selectively eliminated or combined to develop an explanation of the phenomenon under consideration. When only one hypothesis is generated and an attempt is made to demonstrate its correctness, it becomes a 'ruling hypothesis', which dominates the thinking of an investigator and may lead to serious error. The preferred method then is to develop as many explanations of a phenomenon as possible. Through the process of data collection these hypotheses are either modified or eliminated until a solution is developed, or perhaps until multiple explanations or hypotheses are combined to obtain a composite solution or theory.

Figure 2.1 is my attempt to organize a statement of the scientific method that generally follows Beveridge's approach (see Harvey, 1969, p. 34, Haines-Young and Petch, 1986, p. 25). This is a six-step procedure, but unlike Beveridge's the first step is preparation, which involves training and literature review. Without this background it is not possible to recognize a significant problem. In step 3 the

11

problem statement may come directly from the literature, where it has been formulated by another, or it can come from observation of natural phenomena (2a) or by recognition of an anomaly or trend in data collected for another purpose (2b). For example, when a prepared earth scientist views the landscape on aerial photographs or from an aircraft, a great range of landform types, and vegetational and erosional patterns will be observed, and hypotheses can be developed to explain this variability. Also, a vast amount of data has been collected by the Water Resources Division of the Geological Survey primarily for water management purposes, but anyone examining these data will discover water and sediment discharge variations and anomalies that are of scientific and practical importance. Again, hypotheses can be formulated in an attempt to explain these deviations.

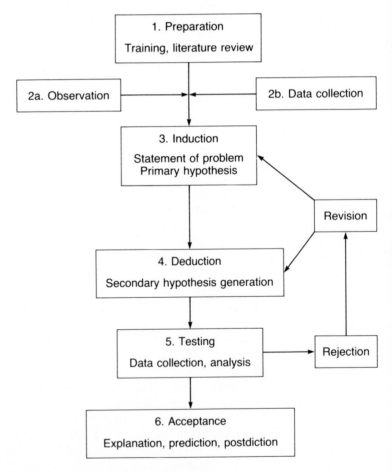

Figure 2.1. An outline of a scientific method.

Methods

From training (1) and experience (2) a problem is recognized by induction (Fig. 2.1), which can be considered to be a primary hypothesis (3). With the statement of the primary hypothesis deductions are made (4), which can be referred to as secondary hypotheses. These can be tested, and this testing will lead to either acceptance or falsification of the primary hypothesis.

Step 5 is the real work of testing hypotheses by field work and data collection, experimentation, etc. Falsification of a hypothesis may lead to its total rejection and failure or, more likely, revision of the hypothesis and renewed testing. Acceptance of the hypothesis will lead to explanation of the problem stated in step 3, and to acceptance because there is a high probability that the hypothesis is correct (6).

Figure 2.2 is an attempt to illustrate steps 4, 5 and 6 of Fig. 2.1. Given a problem (P), hypotheses (H) are generated in order to find an explanation or solution (ES) that is based upon a most probable hypothesis. When only a single ruling hypothesis is developed (*a*), the probability of being wrong is increased, and, of course, the researcher's reputation is at risk if it proves to be wrong. The creation of multiple hypotheses avoids this probability, either in a multiple sequential mode (*b*), with one hypothesis following another as weaknesses are found in each, or in a multiple parallel mode (*c*) with a number of hypotheses being developed and tested simultaneously. Finally, there is the combining of multiple explanations or

(*a*) Ruling hypothesis

(*b*) Multiple sequential hypotheses

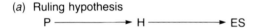

(*c*) Multiple parallel hypotheses

(*d*) Composite hypotheses

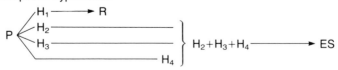

Figure 2.2. The paths to explanation (P = problem; H_1–H_x = hypotheses; H = ruling hypothesis; ES = explanation and solution; R = rejection).

13

hypotheses to develop a composite hypothesis (*d*). In fact, most explanations of complex phenomena will involve a combination of hypotheses, and many great controversies in science end in a compromise.

Part of the difficulty in outlining a scientific method is that there are at least four different levels of scientific activity (Fig. 2.3). Many investigators will enter Fig. 2.1 at step 3 with the problem stated by a supervisor (Fig. 2.3(*b*)). This is a lower level of scientific effort than the recognition of the problem as a result of experience and observation (Fig. 2.3(*a*)). An even lower level of research occurs when the investigator is presented with the hypothesis to be tested (FIg. 2.3(*c*)). The lowest level of effort is when the investigator is instructed to collect and analyze certain types of data. The results are then given to the principal investigator for evaluation. Depending upon what stages of the method one experiences (Fig. 2.1), one's perspective of the method will be affected. That is, someone entering at lower levels may never realize the importance and difficulty of problem recognition and hypothesis generation.

Differential diagnosis

Closely related to the method of multiple working hypotheses (Fig. 2.2) is the process of differential diagnosis in medicine. Harvey and Bordley (1970) deal with this topic, in a book that consists of 1238 pages of information on the diagnosis of various diseases. These authors define differential diagnosis as the *art* of distinguishing one disease from another and of selecting the disease which comes closest to explaining the clinical and laboratory findings. The physician must formulate his hypotheses, when he examines a patient, and these working hypotheses direct the diagnostic process. An outline of the steps in this diagnosis provides another outline of the scientific method (Harvey and Bordley, 1970, p. 7). There are basically three main steps in the diagnosis as follows:

1. Collecting the facts:
   (a) clinical history, (b) physical examination, (c) ancillary examinations, and (d) history of the course of the illness.
2. analysis of the facts:

(a)	(b)	(c)	(d)
Preparation	————	————	————
Observation	————	————	————
Recognition of problem	————	————	————
Statement of hypotheses	Statement of hypotheses	————	————
Data collection	Data collection	Data collection	Data collection
Data analysis	Data analysis	Data analysis	Data analysis
Evaluate hypotheses	Evaluate hypotheses	Evaluate hypotheses	————

Figure 2.3. Outline of four different levels of research.

(a) critically evaluate the collected data, (b) list reliable findings in order of apparent importance, (c) select one or preferably two or three central features of the illness (d) list diseases in which the central features are encountered.

3. Reach a final diagnosis by selecting from the list of diseases either:
    (a) disease which best explains all the facts or (b) several diseases, each of which best explains the facts.
4. Review all the evidence both seemingly positive and seemingly negative with the final diagnosis in mind.

The physician must keep an open mind and be aware of the many diseases that may have symptoms similar to those of the patient. To give some idea of the problems with and complexities of this procedure, Table 2.1 is a list of the causes of diarrhea, something that all readers should be familiar with. We see that the physician can have an abundance of hypotheses for the cause of diarrhea in a particular instance, and, of course, the causes can be multiple (Fig. 2.2). Although there are seven classes of acute diarrhea and six classes of chronic diarrhea with a total of 86 specific cases, the physician will probably give his patient one reason and prescribe for the problem.

A fine example of the use of multiple hypotheses can be obtained from the life work of Ignaz Semmelweis (1861) who used this method several decades before it was described by Gilbert and Chamberlin. Semmelweis was for a period in charge of obstetrics at the Vienna Lying-in Hospital, which was divided into two divisions. All medical students worked with the patients in the First Division and student midwives were assigned to the Second Division. In general treatment of patients was similar in both divisions, but in the First Division the mortality rate was a shocking 10%, whereas in the Second Division it was 3%.

Semmelweis tested several hypotheses to explain this disparity with regard to 1. overcrowding, 2. mode of examination of patients, 3. diet, 4. position of women during labor, 5. ventilation, and 6. cleanliness, and he eliminated all as a cause of the difference in the death rates. He even made the priest, who delivered last rites, be less obvious in the First Division, where, unlike the Second Division, he passed through the ward before reaching the dying patient. None of the changes instituted by Semmelweis were effective in reducing deaths in the First Division.

In 1847 Semmelweis took a short vacation, and upon his return he discovered that his friend Prof. Kolletschka had died with the same symptoms as the women with child-bed fever. The source of the infection was a cut finger received during a post-mortem examination. Semmelweis, in the absence of a germ theory, concluded that his death was caused by 'cadaveric particles, which were introduced into his vascular system'. The next step in his analysis was the assumption that medical students who went directly from autopsies to the hospital introduced the cadaveric particles to the pregnant women. He examined hospital records and

Table 2.1. *Causes of diarrhea (from Harvey and Bordley, 1970, p. 445)*

I. Acute
  A. Intrinsic disease of the gastro-
     intestinal tract
     1. Appendicitis
     2. Diverticulitis
     3. Ischemia of the bowel
     4. Radiation enteritis
     5. Acute ulcerative colitis
     6. Pseudomembranous
        enterocolitis
     7. Partial intestinal obstruc-
        tion (e.g. intussusception)
     8. Acute exacerbation of
        chronic diarrhea
  B. Infections of the intestinal tract
     1. Viral enteritis
     2. Salmonella enteritis
     3. Shigella enteritis
     4. Amebic colitis
     5. Cholera
     6. Staphylococcal entero-
        colitis
     7. Giardiasis
     8. Clostridium welchii
  C. Toxins and poisons
     1. Staphylococcus
     2. Clostridium botulinum
     3. Clostridium welchii
     4. Heavy metals
     5. Mushrooms
     6. Carbon tetrachloride
  D. Drugs
     1. Cholinergic drugs
     2. Ganglionic blocking
        agents
     3. Antibiotics
     4. Colchicine
     5. Digitalis
     6. Iron
       7. Para-amino salicylic acid
       8. Antimalarials
  E. Generalized disorders or
     diseases affecting the intestine
     1. Trichinosis
     2. Malaria
     3. Uremia
     4. Cholemia
     5. Pernicious anemia
     6. Adrenal insufficiency
     7. Hyperthyroidism
     8. Diabetes mellitus
     9. Psittacosis
     10. Weil's disease
     11. Carcinoid syndrome
  F. Acute exacerbation of chronic
     diarrhea
  G. Psychogenic causes

II. Chronic
  A. Intrinsic disease of the gastro-
     intestinal tract
     1. Ulcerative colitis
     2. Diverticulitis
     3. Irritable colon
     4. Colonic neoplasms
     5. Partial intestinal obstruc-
        tion
     6. Inadvertent gastro-
        ileostomy
     7. Regional enteritis
     8. Carcinoma of the stomach
     9. Familial polyposis of the
        colon
  B. Infection of the gastrointestinal
     tract
     1. Amebic colitis
     2. Actinomycosis
     3. Tuberculosis

16

Table 2.1 (*cont.*)

4. Giardiasis	b. Sympathomimetic amine-secreting tumors
5. Strongyloidiasis	
C. Drugs and poisons	
1. Cathartics	c. Zollinger-Ellison syndrome
2. Thyroid	
3. Digitalis	d. Carcinoma of the pancreas
4. Iron	
5. Mercury	e. Carcinoma of the liver
D. Generalized disorders or diseases affecting the intestine	f. Chronic myeloid leukemia
	g. Pancreatic adenoma without gastric hyper-secretion
1. Endocrine and metabolic	
a. Uremic colitis	
b. Hyperthyroidism	h. Nodular carcinoma of thyroid with metastases
c. Addison's disease	
d. Hypervitaminosis D	4. Deficiency diseases
e. Diabetes mellitus	a. Pellagra
f. Cushing's syndrome	b. Pernicious anemia
g. Amyloidosis	5. Other
h. Hypoparthyroidism	a. Cirrhosis of the liver
2. Connnective tissue diseases	b. Chronic cholecystitis
	c. Chronic pancreatitis
a. Systemis Iupus erythematosus	d. Lymphogranuloma venereum
b. Progressive systemic sclerosis	e. Protein-losing enteropathy
c. Polyarteritis'	E. Malabsorption symdrome due to various causes
3. Neoplastic disease	
a. Carcinoid syndrome	F. Psychogenic

determined that the hospital opened in 1784 and from that time until 1841 the mortality rate was about 1 per cent, but after 1841, when autopsies were begun, the mortality rate increased to 8 per cent in 1841, 15.8 per cent in 1842, and 9 per cent in 1843.

The explanation for the disparity in death rates in the two divisions of the hospital was because medical students performed autopsies and went directly from the autopsy to the patients in the First Division. They did not examine patients in the Second Division. The student midwives did not attend autopsies, and therefore, they did not infect their patients in the Second Division.

Semmelweis forced the physicians and students to wash their hands in a solution of chlorine, and the mortality rates became comparable in the two divisions. Semmelweis' reasoning was correct, but unfortunately he could not convince the medical establishment, and his method was not generally followed, which was a great tragedy. Nevertheless, Semmelweis provides us with an excellent example of multiple working hypotheses, and the generation of a new hypothesis as new information becomes available, which appears to be a combination of methods (*b*) and (*c*) of Fig. 2.2.

Method in earth sciences

In order to function under the conditions imposed by their science, earth scientists have used the 'principle of uniformitarianism' (uniformity), and reasoning by analogy. Geikie (1905, p. 299) described uniformity as 'the present is the key to the past'. Europeans prefer the term actualism (actualisme, Aktulismus), which refers to explanation, postdiction and prediction based on the understanding of present processes (Jong, 1966; Rutten, 1971). There is currently considerable controversy concerning uniformitarianism (Albritton 1963, 1967; Gould, 1965, 1984), which has either been lauded as a basic principle of geology or as a worthless concept. For example, Challinor (1968) is of the opinion that the present tells us very little about the past and that, in fact, the past tells us about the present (Kitts, 1966, p. 143). The concept of uniformitarianism has also been greatly misinterpreted, and Shea (1982) concludes that 'most geologists do not understand the nature and correct meaning of what is said to be the basic principle of their science'.

The controversy surrounding the definition of uniformity is somewhat analogous to the reinterpretation of the US Constitution in the light of modern conditions. The fact that neither is defined rigorously leads to conflicting opinions. In order to remove this difficulty the restricted definition of uniformity as used by physicists, chemists and philosophers, can be employed. It is simply the assumption that natural laws are permanent; that is, under the same conditions a given cause will always produce the same results. Therefore, 'Regularities in the sequence of type of event found in any restricted region of space and time will continue to hold good in all regions of space and time.' (Harré, 1960, p. 117). This concept is necessary for extrapolation, from the present to the past and to the future.

Nothing stated above excludes the rare event such as meteor impact (Kyte *et al.*, 1988) and massive flooding as in the Channeled Scablands (Baker, 1973). They are improbable events that can be explained rationally. Such events probably can be safely eliminated from short-term predictions, but not for geologic time (Gretener, 1967, 1984), and they do not violate uniformity.

Analogy

Explanation of past events and prediction of future events (application of uniformity to the past and future) frequently requires reasoning by analogy (Gilbert,
1896; Leopold and Langbein, 1962), which is the recognition of similarity among
different things. Analogy is indeed the basis of explanation and extrapolation for
geologists, which is why we need experience and need to see as much as we can
of different existing (actual) conditions so that comparisons can be made among a
variety of geologic phenomena (Jong, 1966, p. 238). For example, planetary geologists use analogic arguments to explain features that cannot be inspected 'in the
field' (e.g. Schumm, 1970; Baker and Milton, 1974).

Analogy was recognized as a significant component of scientific discovery by
Francis Bacon, Booke, Kepler, Mach, Maxwell, Poincaré, Leatherdale (1974,
pp. 4 and 13), and Medawar (1984, p. 109). Analogy in science is based on the
assumption that if two things are alike in some respects then they must be alike in
others. Frequently the similarity among things leads to a flash of insight and an
inductive leap that solves a problem. Needless to say, many similarities may be
misleading, and the recognition of analogous conditions is not sufficient to estab-
lish more than a working hypothesis. Indeed, Von Bertalanffy (1952, p. 200) insists
that analogies are worthless but that homologies are useful because they involve
more than superficial similarities, and they involve similarities in both structure
and function. Extrapolation will be strengthened by use of homology because
analogy is like a metaphor, whereas homology is like a synonym. Homology is a
higher level of comparison than analogy. The discussion of the use of analogy and
metaphor in science is extensive (e.g. Livingstone and Harrison, 1981).

A problem with using analogy in geomorphology is that, where erosion is
moderate or slow, the modern landscape may be dominated by past events which
greatly complicate the analogy (Douglas, 1980). Areas subjected to continental
glaciation retain that influence, and on a different scale the patterns of some
modern rivers reflect the influences of active tectonics on valley-floor slope
(Adams, 1980; Burnett and Schumm, 1983). The question, then, becomes one of
the validity of using modern analogy as a basis for extrapolation. There really is no
other option, and the actualistic approach is mandatory. In fact, the concept of
actualism can be modified and expanded to include the future as follows: the
present provides insights into the past, and it influences future geologic processes
and events, and the past influences both present and future geologic processes and
events (Fig. 1.2).

Multiple hypotheses

Although, unlike other sciences, there are no books describing the scientific
method as applied to geologic problems, there are several papers that stress the

multiple hypothesis approach (Fig. 2.2), and in each case the procedure is illustrated by geomorphic examples (Gilbert, 1886, 1896; Chamberlin, 1890, 1987). Because these papers are of such importance they will be reviewed here.

Gilbert:

Gilbert (1886) characterizes his method of research as follows:

1. selective and concentrated observation;
2. empirical classification and grouping of facts;
3. development of a hypothesis by induction to explain observations;
4. testing the validity of hypothesis and revising until explanation is satisfactory.

An important contribution of Gilbert, which anticipates Chamberlin's 1890 paper on multiple working hypotheses, is his statement that the investigation 'is not restricted to the employment of one hypothesis at a time. There is indeed an advantage to entertaining several at once . . . ' (Gilbert, 1886, p. 286). He warns of the dangers of a single ruling hypothesis and stresses that the development of hypotheses depends on analogy.

The example that Gilbert uses to illustrate his method is the variable height of a Pleistocene shoreline of Lake Bonneville, Utah. A previous worker had determined the elevation of the shoreline at one location and assumed that it was horizontal. Gilbert measured the elevation at two locations and found that the elevations were not the same, and, therefore, the shoreline was not horizontal. He first assumed that 'a gentle undulatory movement of the crust' or folding was the cause, but he then observed a fault between the location of the measurements and this additional hypothesis, faulting, was added as a probable explanation. Gilbert carried out additional surveys, as part of his general study of the geology of Lake Bonneville, and he found that the displacement along the fault was not adequate to explain the difference in elevation of the shoreline, therefore, the faulting hypothesis was rejected.

As more information became available, he discovered that the maximum displacement of the shoreline occurred over the center of Lake Bonneville, and he proposed the deformation was somehow related to the drying up of Lake Bonneville to form Great Salt Lake. He eventually suggested that the warping was the result of what we now call isostatic adjustment of the crust to the removal of the load produced by evaporation of the water, but in order to prove this hypothesis additional surveying of the complete shoreline and surveys of other Pleistocene lakes were required. If all showed similar shoreline deformation the isostatic hypothesis would be strongly supported, and of course, it was (Crittenden, 1963), and the folding and faulting hypothesis was abandoned. Gilbert apparently used the method of multiple sequential hypotheses (Fig. 2.2(*b*)).

Chamberlin:

Shortly after Gilbert's paper appeared the famous paper of Chamberlin (1890) was published. This paper has received much attention during the last 100 years (Laudan, 1980), and Platt (1964, p. 350) describes the method of multiple working hypothesis as a 'great intellectual invention'. In any case, the paper was published twice by Chamberlin and reprinted at least three times.

Chamberlin's objective was to stress the need for multiple hypothesis generation, a procedure used by Gilbert (1886) and Semmelweis (1861) earlier. Chamberlin stressed the danger of a ruling hypothesis (Fig. 2.2(*a*)) as follows:

> The moment one has offered an original explanation for a phenomenon which seems satisfactory, that moment affection for his intellectual child springs into existence; and as the explanation grows into a definite theory, his parental affections cluster about his intellectual offspring, and it grows more and more dear to him, so that, while he holds it seemingly tentative, it is still lovingly tentative, and not impartially tentative. So soon as this parental affection takes possession of the mind, there is a rapid passage to the adoption of the theory. There is an unconscious selection and magnifying of the phenomena that fall into harmony with the theory and support it, and an unconscious neglect of those that fail of coincidence. The mind lingers with pleasure upon the facts that fall happily into the embrace of the theory, and feels a natural coldness toward those that seem refractory. Instinctively there is a special searching-out of phenomena that support it, for the mind is led by its desires. There springs up, also, an unconscious pressing of the theory to make it fit the facts, and a pressing of facts to make them fit the theory. When these biasing tendencies set in, the mind rapidly degenerates into the partiality of paternalism. The search for facts, the observation of phenomena and their interpretation, are all dominated by affection for the favored theory until it appears to its author or its advocate to have been overwhelmingly established. The theory then rapidly rises to the ruling position, and investigation, observation, and interpretation are controlled and directed by it. From an unduly favored child, it readily becomes master, and leads its author whithersoever it will. The subsequent history of that mind in respect of that theme is but the progressive dominance of a ruling idea.

Even when the investigator tries to be unbiased and to test a single hypothesis objectively, it is easy for it to become a ruling hypothesis. The solution is clear; multiple hypotheses must be proposed (Fig. 2.2(*c*)) and

> The investigator thus becomes the parent of a family of hypotheses: and, by his parental relation to all, he is forbidden to fasten his affections unduly upon any one. Having thus neutralized the partialities of

his emotional nature, he proceeds with a certain natural and enforced erectness of mental attitude to the investigation, knowing well that some of his intellectual children will die before maturity, yet feeling that several of them may survive the results of final investigation since it is often the outcome of inquiry that several causes are found to be involved instead of a single one. In following a single hypothesis, the mind is presumably led to a single explanatory conception. But an adequate explanation often involves the coordination of several agencies, which enter into the combined result in varying proportions. The true explanation is therefore necessarily complex. Such complex explanations of phenomena are specially encouraged by the method of multiple hypotheses, and constitute one of its chief merits. We are so prone to attribute a phenomenon to a single cause, that, when we find an agency present, we are liable to rest satisfied therewith, and fail to recognize that it is but one factor, and perchance a minor factor, in the accomplishment of the total result.

Not only is the method of multiple hypotheses a significant contribution, but the resulting statement by Chamberlin that complex or composite explanations of a feature are probable is equally significant (Fig. 2.2(*d*)), but this point has been ignored in the discussions of Chamberlin's contributions. Chamberlin uses as an illustration the deep basins of the Great Lakes, which had been previously explained as river valleys, glacial excavations, or as the result of downwarping. Chamberlin states that all of these explanations are partly correct.

The problem therefore, is the determination not only of the participation, but of the measure and the extent, of each of these agencies in the production of the complex result. This is not likely to be accomplished by one whose working hypothesis is pre-glacial erosion, or glacial erosion, or crust deformation, but by one whose staff of working hypotheses embraces all of these and any other agency which can be rationally conceived to have taken part in the phenomena.

A final quotation from the 1897 (p. 165) paper provides an adequate summary of Chamberlin's views.

The studies of the geologist are peculiarly complex. It is rare that his problem is a simple unitary phenomenon explicable by a single simple cause. Even when it happens to be so in a given instance, or at a given stage of work, the subject is quite sure, if pursued broadly, to grade into some complication or undergo some transition. He must therefore ever be on the alert for mutations and for the insidious entrance of new factors. If, therefore, there are any advantages in any field in being armed with a full panoply of working hypotheses and in habitually employing them, it is doubtless the field of the geologist.

Methods

Gilbert:

Gilbert's (1896) later paper is his attempt to illustrate in some detail the method of multiple hypotheses, which, ironically, in this case led to an incorrect conclusion. Gilbert again stresses the role of analogic reasoning in hypothesis generation.

The example he selects to illustrate the approach is his investigation of a large crater southeast of Flagstaff, Arizona called Coon Butte and now known as Meteor Crater. This large approximately circular crater is 'a few thousand feet broad and a few hundred feet deep' with a rim rising above the plain of Kaibab limestone. Although the San Francisco volcanic field is nearby, the crater is comprised of limestone and sandstone. However, meteor fragments are found nearby. Four hypotheses for the crater and iron fragments were proposed by Gilbert and others as follows:

1. it is a volcanic explosion crater from which the fragments were ejected;
2. it is a breached dome over a laccolithic intrusion at depth;
3. it is a crater formed by steam explosion;
4. it is a meteor crater.

A magnetic survey by Gilbert showed that a meteor was not buried in the crater, and a calculation of the volume of ejected material revealed that the ejected material would fill the crater leaving no space for a meteor (Fig. 2.4(*a*)). As there are no igneous rocks associated with the crater, all hypotheses are rejected except that of a phreatic explosion. Elsewhere in the world similar appearing craters (Maars) were formed by explosions, and a nearby volcanic field indicated that a source of geothermal energy existed (Fig. 2.4(*b*)); therefore, Gilbert accepted the volcanic origin of Meteor Crater. If he had known that the impact of the meteor at high velocity causes an explosion (Cooper, 1977) and that a relatively small meteor could have formed Meteor Crater, his conclusion undoubtedly would have been different.

The example of Gilbert's study of Meteor Crater shows that in spite of a determined effort to use multiple hypotheses erroneous conclusions can result. This is especially true when the records are fragmentary, as in most studies of structural geology, stratigraphy, sedimentation and historical geomorphology.

Douglas Johnson (1933, 1940) elaborated on Gilbert's and Chamberlin's suggestions and described a seven-step method of scientific analysis that differs little from others discussed above. More recently Platt (1964) urges the systematic application of multiple hypotheses with careful design of crucial experiments that will exclude some hypotheses. Mosley and Zimpfer (1976) stress, in addition, that composite explanations are often required when dealing with complex geomorphic systems. They use an analogy with analysis of variance and conclude that an explanation of a complex phenomenon, such as river meandering, requires partial answers from geology, geomorphology, hydraulics, geotechnical engineering, and hydrology (Fig. 2.2(*d*)).

If a ruling hypothesis proves correct then all credit to its parent. However, if it is in error the consequences to a career may be unfortunate and significant. The multiple hypothesis approach, although difficult, is an ideal that should be the goal of the earth scientist.

Analogy with scientific paper

It may be of some value to consider that a well written scientific paper is analogous to the scientific method. Even though the scientific paper does not describe the mistakes made and the false starts during the performance of research, it provides an idealized example of how science should be carried out (Table 2.2). For example, the introduction of the scientific paper includes a statement of the problem, a review of relevant literature, and a statement of the objectives of the specific study. In research this is the period of preparation, problem identification, and hypothesis generation (Fig. 2.1). The second part of a paper is a description of the procedure followed, which in research involves development of a work plan to be followed for data collection and analysis. The third part of the paper is the presentation of the results of data collection and analysis, which is obviously the next step

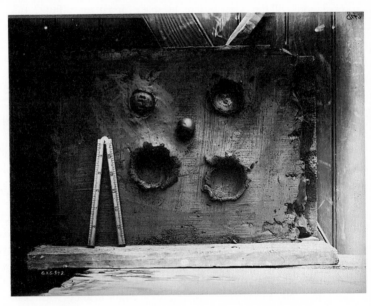

Figure 2.4(*a*). G. K. Gilbert's experiments on crater formation by impact. Balls of clay were thrown against a slab of clay. Different size craters are due to different impact velocities. Note that the clay ball occupies a large part of the crater volume. (G. K. Gilbert photograph 842, circa 1891, US Geological Survey Photolibrary, Denver.)

in the research program. In the paper the final conclusions are presented, but in reality only if we are very lucky is the original hypothesis supported, and it may be necessary to modify the hypothesis and to generate other hypotheses which will then be tested.

Unfortunately, as noted above, the scientific literature gives little hint of the trials of scientific research, and the papers that are produced only describe successful methods, successful data analysis and conclusions which support the original or modified hypotheses. Of course, this is because of page limitations and the inability of editors to permit a discussion of the full amount of work done by a scientist, which would include the normal confusion and disappointments inherent in research. Therefore, the despair of success frequently experienced by a young researcher may seem to him unique, whereas it is a common attribute of research. The senior scientist in such cases must advise the young investigator to proceed with his work because the hypothesis is worthy of investigation and because the end results will be of value, even in the disappointing circumstance that they do not support the stated hypotheses. Unfortunately, the scientific literature gives a very misleading perspective of science. It isn't as easy as the papers suggest.

Fig. 2.4(*b*). G. K. Gilbert contemplating cinder cones near Flagstaff, Arizona. (G. K. Gilbert photograph 801, 1891, US Geological Survey Photolibrary, Denver.)

Table 2.2. *Comparison of the scientific method to the outline of a scientific paper*

Scientific paper	Scientific method
1 Introduction	Preparation
Statement of problem	Statement of problem
Literature review	Literature review
Statement of objectives	Statement of objectives
2 Procedure	Work plan
Date collection	Data collection
3 Results	Data analysis
Data analysis	Evaluation of results
Results of analysis	
4 Summary and conclusions	Accept, reject or modify hypothesis

CRITICISM OF THE SCIENTIFIC METHOD

Although there is agreement that a general scientific method or approach exists, there are critics who convincingly argue that the method is only as powerful as the objectivity of the individual using it. Although the multiple hypothesis approach is obviously the way to proceed, the difficulties of eliminating bias are nicely stated by N. L. Bowen (1948, p. 79) as the 'method of multiple prejudices'. This is in contrast to multiple working hypotheses where

> Chamberlin thus had in mind a happy situation where an individual investigator diligently sought all reasonable processes that might lead to an observed relation, carefully considered the full consequences of each process envisioned, impartially compared these deduced consequences with the re-examined facts and thus reached a conclusion as to the probable process or group of processes that were operative. 'Tis a consummation devoutly to be wished; yet one . . . cannot fail to wonder whether any individual is capable of such detachment. Each of us is, of course, quite sure that he himself has done just what Chamberlin recommends, but he is equally sure that the other fellow has done no such thing . . . That fictional character, the impartial observer, would probably say that we are all nursing pet prejudices.

Perhaps the severest criticism of the concept of a scientific method is presented by Kuhn (1970) and Feyerabend (1975, 1978). They make a valid point that, depending upon a person's social status, economic background, and training, and the prevailing social and scientific climate, it is possible to view scientific problems in different ways (see Haines-Young and Petch, 1986). For example, if one's geomorphic training was in an area of a humid climate and low relief, a low-energy system, one would conclude that very little has happened since the Pleistocene, and for a long time, British geomorphologists did not involve themselves in studies of erosion process and rates of change because the senior people were conditioned to believe that very little had happened since the last glaciation in the British Isles. During this time American geomorphologists were carrying out numerous studies of hillslopes and rivers, which provided information on rates of change and the response of the landscape to climatic change.

Oliver Sacks (1987, pp. 92–101) provides a neurological example. Tourette's syndrome is characterized by an excess of nervous energy and by tics, jerks, grimaces, noises, etc. Tourette described this condition in 1885, but in this century the syndrome seemed to have disappeared, and it was rarely diagnosed in the first half of this century. According to Sacks, some physicians regarded it as a 'product of Tourette's colorful imagination', but Sacks diagnosed a case in 1971. The next day 'without especially looking' he saw two more in the streets of New York. He was prepared to see rather than to ignore.

We tend to see what we are trained to see. This was brought home to the author, a geomorphologist, in a very forceful way while travelling in a car with a soil scientist, a structural geologist, and a botanist. All four of us were viewing the same landscape. A discussion with regard to a particular location revealed that the botanist was observing the vegetation, but he did not look through the vegetation to see the variations of soil, which were obvious to the soil scientist. I was looking at erosion features on a hillslope, when the structural geologist pointed out that the hill was an anticline. I had looked through the vegetation and the soil to the erosion features but ignored the geology. The structural geologist on the other hand had ignored the vegetation, soil and geomorphology to observe the very obvious structural folding that the other three had ignored. An individual cannot be criticised for the bias of his profession, but one should be aware of it.

A fictitious example of interpretation gone awry as a result of bias is provided by David McCaulay (1979) in his amusing book, *Motel of the mysteries*. In the year 4000 AD a twentieth-century motel is excavated by an archaeologist who was clearly conditioned by his study of early Egyptian civilization. His religious interpretation of the objects found in an average motel room indicate how badly things can go wrong if one does not maintain objectivity. For example, his interpretation of a flush toilet was that it was a sacred urn. During a burial ceremony 'The ranking celebrant, kneeling before the urn, would chant into it while water from the sacred spring flowed in to mix with sheets of Sacred Parchment.' Everything in the bedroom was oriented toward the television, therefore it was the high altar. The purpose of the multiple hypotheses approach is to prevent this type of error.

A bias of training becomes even more obvious when the same situation is viewed by scientists and engineers. For example, there can be three approaches to incised channel (gully, arroyo, entrenched stream) stabilization. The geomorphic approach can be to 'let it go,' as it will eventually reach a new condition of relative stability. The engineering approach will normally be to use a variety of structures to control the problem. There can also be an intermediate 'rational' approach that incorporates geology, geomorphology, and engineering. This approach uses engineering structures only at sites selected by a careful study that has identified the different stages of incised-channel change and that can provide the basis for selecting where engineering structures will be effective and where they are redundant or useless (Schumm *et al.*, 1984). This is similar to Mackin's (1963) rational approach to scientific problems where understanding is stressed rather than facile acceptance of purely statistical results (Klemés, 1985).

The problem is, of course, that the same field situation or 'the same set of experimental data can often be interpreted in more than one way – which is why the history of science echoes with as many venomous controversies as the history of literary criticism'. 'The data may be "hard" . . . but what you read into them is another matter' (Koestler, 1978, p. 153).

Resistance to change

There is also an apparent reluctance to accept new ideas in science, which suggests that scientists may be less objective than generally assumed (Barber, 1961) and in fact some unpublished notes of T. H. Huxley reflect this fact (Bibby, 1960, p. 77; Gregory, 1985, p. 10; Stoddart, 1986, p. 16). In the following I have substituted the word theory for Huxley's word novelty:

1. Just after publication – The theory is absurd and subversive of religion and morality. The propounder both fool and knave.
2. 20 years later – The theory is absolute Truth and will yield a full and satisfactory explanation of things in general. The propounder a man of sublime genius and perfect virtue.
3. 40 years later – The theory will not explain things in general after all and therefore is a wretched failure. The propounder a very ordinary person advertised by a clique.
4. 100 years later – The theory is a mixture of truth and error. It explains as much as could reasonably be expected. The propounder worthy of all honor in spite of his share of human frailties, as one who has added to the permanent possessions of science.

The long delay in the acceptance of the concepts of plate tectonics (continental drift) and the rejection of the clear evidence of the great Pleistocene floods that so modified the Scabland landscape of eastern Washington should not, however, be used as evidence that scientists are not objective, as it is difficult to accept new hypotheses when they do not have what appears at the time to be a physical basis. What was the mechanism that moved continents? Where did the water originate that caused the floods? In the latter case, when glacial lake Missoula in western Montana was discovered and linked to the megafloods, the origins of the Channeled Scablands by regional flooding was accepted (Bretz, 1969; Baker,

1973). Hence, not all resistance to change is irrational. The explanation must be related to physical reality.

However, Giere (1988, p. 292) makes the suggestion that the 'virtues of multiple working hypotheses tend to be advanced by scientists attempting to gain a hearing for a new view', and these virtues are minimized by scientists who are defending an established position. This suggests that older scientists are less open to new ideas. Grinnell (1987, p. 292) however, expresses a more charitable view that scientists who are deeply involved in a problem know a great deal about it, and they can legitimately raise objections to new hypotheses. This indeed, makes the younger generation of investigators seem to be more open-minded, but the end result of such a controversy may be that 'older hypotheses are not disproved so much as they are replaced by newer ones' (Grinnell, 1987, p. 39).

Falsification

In order to overcome prejudices and conditioning the 'critical rationalist approach' of Popper (1968) produces a conscious attempt to disprove or falsify a hypothesis. That is, we should attempt to disprove rather than verify our hypotheses. One may argue that we do falsify hypotheses when the data fail to support them, but it is the negative and critical state of mind that is cultivated in the falsification approach that is encouraged. Although philosophically this may be advantageous and a more objective approach, it is difficult to imagine most scientists struggling to disprove their ideas, where, as is the case in earth science, supporting or disproving data may be so limited.

Popper argues that it is impossible to prove the correctness of a hypothesis, but one can disprove it by one failure. We can take hope in the assumption that science is 'self-correcting'. That is, error or fraud will be revealed with time. Popper's approach is certainly different from what has been described as the scientific method, but it could lead to identification of new data that are needed or additional observations that need to be made that will either support a hypothesis or destroy it.

An example of the falsification approach is provided by a medical situation described by Rouché (1984) in which the disease rikettsial pox was diagnosed and the treatment established by Doctor Benjamin Shankman. In February, 1946 a young boy who had a spotty rash, high fever and considerable joint pains was examined by Dr Shankman. The doctor was unable to diagnose this disease. The boy was sufficiently ill to be placed in the hospital and he was treated with penicillin for infection, codeine for pain and aspirin for fever. After approximately five days the boy recovered. Shortly afterwards a woman with the same symptoms was admitted to the hospital by Dr Shankman. He wondered whether penicillin really resulted in the cure of the boy, and he modified his treatment by prescribing sulfonamide instead of penicillin. The woman recovered. A third patient arrived with the same symptoms, and Dr Shankman questioned the need for antibiotics.

He treated the woman only with codeine and aspirin, and the woman recovered. Dr Shankman's conclusion was that the disease would run its course in a few days and that the patient would recover without any special treatment other than that required to reduce joint pain and fever (codeine and aspirin). Although in the first instance the treatment was successful, Dr Shankman was not convinced that the treatment was necessary, and he falsified his hypothesis that an antibiotic was required. It was later determined that the disease was transmitted by mites living on mice in a large apartment complex in New York. In this case, the diagnosis of the disease was correctly established long after the 'cure' was found by Dr Shankman and, therefore, his falsification procedure was applied to treatment rather than to the cause of the disease.

Probability and quality of a hypothesis

Strahler (1987), in his excellent book on the evolution–creation controversy takes issue with Popper's approach and suggests that if it is not possible to prove a hypothesis then it is not possible to disprove or falsify it. Rather we should think of the probability of the hypothesis being false or true.

He states his case as follows:

> To summarize the concept of quality of a scientific hypothesis and to give it graphic expression that may make it easier to grasp, I have drawn a . . . 'ladder of excellence' [Fig. 2.5] on which each 'rung' denotes a tenfold increase in quality. We must take into account two complimentary statements of probability:
>
> P_T: Probability that the hypothesis is true.
>
> P_F: Probability that the hypothesis is false.
>
> The sum of these two probabilities is unity; therefore, as P_T increases, P_F decreases. However, we must keep in mind that neither value can actually reach either unity or zero – those limits can only be approached, so our ladder has no upper or lower end. Gambling odds on the hypothesis actually being a true statement of nature are given as the ratio of P_T to P_F. I have suggested adjectives to describe the quality of the hypothesis, but these are quite subjective because, as with people in general, scientists differ among themselves as to the risks they are willing to take in a given situation. At some point high on the ladder, the hypothesis may take on the status of a law of science.

Certainly in the earth sciences all except the simplest hypotheses will have a degree of uncertainty, and Stahler's approach makes a good deal of sense. The problem, of course, is to determine the probabilities of an accurate extrapolation.

Beveridge (1980, p. 58) is also very critical of the falsification approach. He claims that many important scientific advances have been made in spite of apparent falsification, and it was only the investigator's persistence in the face of negative results that led to a solution. Most of the time a scientist is 'trying to achieve some objective, or to understand some phenomenon, not to refute some hypothesis. Antibiotics and the helical structure of DNA were not discovered by trying to disprove a theory'. He would agree with Strahler that it is difficult to design an experiment that gives a clearly negative result, and therefore, most results are expressed, as Strahler suggests, in statistical terms. It seems that falsification of a hypothesis is as uncertain as its confirmation (Grinnell, 1987).

Popper would agree with these criticisms because he is aware that 'no conclusive disproof of a theory can ever be produced' (Popper, 1968, p. 50) and 'The game of science is, in principle, without end. He who decides one day that scientific statements do not call for further test, and that they can be regarded as finally verified, retires from the game' (Popper, 1968, p. 53). Indeed, Chapter 6 of his book deals with degrees of falsifiability. Furthermore, he makes a modern statement of multiple working hypotheses as follows: 'According to my proposal, what characterizes the empirical method is its manner of exposing to falsification, in every conceivable way, the system to be tested. Its aim is not to save the lives of vulnerable systems but, on the contrary, to select the one which is by comparison

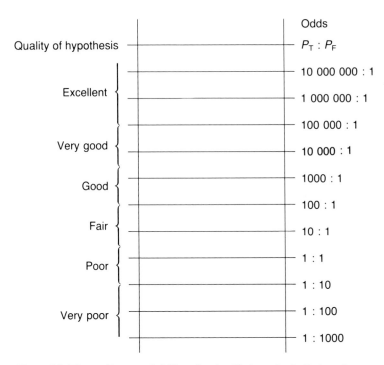

Figure 2.5. The quality or probability of a scientific hypothesis. P_T is probability of being true or of enjoying long survival. P_F is probability of being false or of being short lived (from Strahler, 1987).

the fittest, by exposing them all to the fiercest struggle for survival' (Popper, 1968, p. 42). Substitute in the above the word scientific for empirical and the word hypotheses for system.

Pseudoscience and the scientific approach

Obviously there is no single 'scientific method', but all scientists attempt to approach a problem in such a way that their conclusions are tested. Bunge (1984) contrasted the characteristics of science and pseudoscience, and this may also help to identify a general approach that is scientific. He describes both science and pseudoscience as cognitive fields which are 'a sector of human activity aiming at gaining, diffusing, or utilizing knowledge of some kind, whether this knowledge be true or false.' The cognitive fields can be divided into research fields (science) and belief fields (Fig. 2.6). A research field is continually changing, as a result of research results, but a belief field changes only as a result of controversy, brute force or revelation. Bunge then compares the attitudes and activities of the scientist and pseudoscientist (Table 2.3). The scientist attempts to function with an open mind and uses certain scientific methods (Table 2.3, items 3, 5, 6, 9–14) and a scientific state of mind or a scientific approach (Table 2.3, items 1, 2, 4, 7, 8, 15). The approach need not include a determined attempt to falsify, but it must include a consideration and testing of other explanations and a rational approach to the problem.

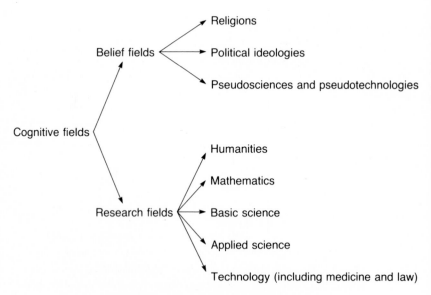

Figure 2.6. Cognitive fields (from Bunge, 1984).

Table 2.3. *Attitudes and activities of scientists (from Bunge, 1984)*

1 Admits own ignorance, hence need for more research
2 Finds own field difficult and full of holes
3 Advances by posing and solving new problems
4 Welcomes new hypotheses and methods
5 Proposes and tries out new hypotheses
6 Attempts to find or apply laws
7 Cherishes the unity of science
8 Relies on logic
9 Uses mathematics
10 Gathers or uses data, particularly quantitative ones
11 Looks for counter-examples
12 Invents or applies objective checking procedures
13 Settles disputes by experiment or computation
14 Updates own information
15 Seeks critical comments from others

DISCUSSION

One may conclude that although there are variations in scientific methods, there is a general scientific approach (Table 2.3) that should lead us on the path to explanation and understanding. Part of the difficulty is the different approaches used in applied and basic research and, of course, the great variety of scientific specialties each with its own problems and techniques.

Although I have attempted to illustrate some approaches to explanation in Fig. 2.2, as based upon the writings of Gilbert and Chamberlin, Fig. 2.7 contains what I perceive to be the probable course of a scientific investigation from recognition of a problem to its solution. This sequence is a combination of all of the paths to explanation of Fig. 2.2. For me, at least, it is difficult to think of more than three or four working hypotheses to explain a problem such as P_1 (H_1, H_2, H_3). As data are obtained, hypotheses can be rejected (H_1, H_3) as being improbable explanations of P_1 but as data are collected, new hypotheses are generated (H_4) and new problems recognized (P_2). Depending upon the importance of the problem, P_1 can be abandoned and P_2 can be followed. Similarly, as the data are analysed, the result will usually cause the abandonment of hypotheses (H_4) and the development of new hypotheses (H_5) and the recognition of new problems (P_3). The final solution will probably be a combination of two or more hypotheses (H_2, H_5) that lead to a composite explanation of the phenomenon of concern (ES) and perhaps recognition of new problems.

The procedure outlined in Fig. 2.7 is obviously a combination of the working hypotheses concepts of Gilbert and Chamberlin and the falsification procedure of Popper. The end result is a most probable composite hypothesis that is the result of a combination of hypotheses.

The warning by Gilbert, Chamberlin and Popper against a ruling hypothesis cannot be too strongly supported. An impressive example is provided by the auto-biographical account of Susan Blackmore's (1986, p. 8) efforts to demonstrate the validity of the paranormal, extrasensory perception; she was 'hooked' by para-psychology because 'Parapsychology has everything a hook needs. It is mysterious and alluring. It has just enough "scientific" evidence to provide bait, while at the same time it is rejected by most orthodox scientists, the inspiration for a crusading spirit to shout "I'll show them". And that is, I suppose, what I wanted to do.' And she did for years until finally convinced by continuing negative results to question the ruling hypothesis. Obviously Blackmore was operating in a 'belief field'. Of greater consequence is the ruling hypothesis in politics, political science and law where individuals and societies are damaged by it.

The scientist is somewhat in the situation of a parent. Each child and each scientific problem is different, and clearly a better job would be done if in both cases we could do it again. In both cases methods will vary, but a consistent approach is mandatory (Table 2.3, Fig. 2.7).

In addition to the basic scientific problems being different there are specific pro-cedural problems that may be encountered in the development of explanations of phenomena and in the extrapolation of research findings to analogous and homol-ogous situations. Ten of these problems will be discussed in Chapter 3.

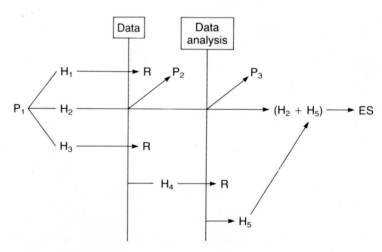

Figure 2.7. The path to explanation (P = problem; H = hypothesis; ES = explanation and solution; R = rejection).

3 Problems of explanation and extrapolation

Too often I have been able to little more than indicate the difficulties likely to be met – yet merely to be forewarned is often a help.
Beveridge, 1957, p. x

The range of problems encountered during research require ingenuity, innovation and persistence. As noted earlier all investigators encounter difficulties during research. In fact, researchers frequently are despondent because 'things are not working right'. Of course, that is the nature of research and discovery. We can be misled by the scientific literature, which always portrays research as being successful. Failures are not publishable, and only infrequently can the steps to discovery be published, blemishes and all. Therefore, it seems worthwhile to discuss the problems that confront the earth scientist. These problems may help to explain the great differences of opinion concerning the scientific method, and the reason that extrapolation is frequently difficult.

There are ten problems associated with using modern conditions as a basis for extrapolation. They are defined as follows: 1. *Time*, involving both absolute duration and relative time spans; 2. *Space*, involving scale and size; 3. *Location*, the site of concern within a natural system; 4. *Convergence* (equifinality), the production of similar results from different processes and causes; 5. *Divergence*, the production of different results from similar processes and causes; 6. *Efficiency*, the variable efficiency and work accomplished by a process; 7. *Multiplicity*, the multiple explanations that combine to influence and cause natural phenomena; 8. *Singularity*, the natural variability among like things; 9. *Sensitivity*, the susceptibility of a system to change; and 10. *Complexity*, the complex behavior of a system that has been subject to altered conditions. Thornes (1983) points out that at least five of the problems, time, convergence, divergence, complexity and sensitivity must be considered during paleoenvironmental reconstruction. Table 3.1 presents two innocuous mnemonic aids to memorization of the ten problems. More effective but inappropriate here would be a cruder phrase which would remain in the mind; create your own.

The ten words that describe each problem could be considered to be an attempt to use 'name magic'. This is an anthropologist's term for the attempt of a primitive people to control a phenomenon by assigning it a name. If we give something a name we often think that we understand it (Weinberg and Weinberg, 1979, p. 313). The scientific literature reveals that not only primitive people use this technique.

Although I view each problem as being distinct and different, there is obvious

35

Table. 3.1. *Mnemonic aids for the memorization of the ten problems*

1 The	—	Time	—	The
2 Sisters	—	Space	—	Scientist
3 Like	—	Location	—	Likes
4 Candy	—	Convergence	—	Contemplation
5 Do	—	Divergence	—	Do
6 Send	—	Singularity	—	Senior
7 Sisters	—	Sensitivity	—	Scientists
8 Candy	—	Complexity	—	Contemplate
9 Every	—	Efficiency	—	Everything
10 Monday	—	Multiplicity	—	Methodically?

overlap and interaction among them. For example, problems of time and space are closely associated. Nevertheless, the examples should demonstrate that each problem is a reality, and indeed, each has been found to be a valid problem that can create significant difficulties during an investigation.

The problems can be grouped into three broad classes that are based upon 1. scale and place, 2. cause and effect and 3. system response. Time, space and location are problems associated with establishment of boundaries for a study and data collection. These three problems are different in type from convergence, divergence, efficiency and multiplicity, which are involved with process. Finally, singularity, sensitivity and complexity are problems that involve system characteristics and response to change.

PROBLEMS OF SCALE AND PLACE

Time

Definition

The problem of time is the most difficult for discussion because time cannot be seen, and it is difficult to define. Time is really no different from other dimensions such as length, but much has been written about it and what it is (Zwart, 1976; R. Morris, 1985). Time for our purpose is what is measured by a clock. It is the measure of change in an earth system that is measured in seconds, minutes, and years just as length is measured in millimeters, centimeters and meters. Time, therefore, is a means of measuring change, and it can be viewed as an index of the rate of energy expenditure, work done, or change of entropy. These variables cannot be measured directly during geologic time, but time can be used as a surrogate for these variables just as drainage basin area can be used as a surrogate for water discharge from ungaged watersheds.

Statement of problem

There are two aspects to the problem of time. The first is availability, which is the specific length or period of time involved in data collection, which is usually too short. Indeed, one of the first problems that arose concerning time in geology was its availability. However, the discovery of radioactivity and radioactive decay, as a source of energy, gave geologists a method for measuring the time needed for the development of geologic phenomena (Burchfield, 1975). Lord Kelvin was proved wrong in his attempt to keep earth history short, although that problem has not been resolved to the satisfaction of creationists (Strahler, 1987).

The second problem with time is that we deal with physical systems that operate over varying time spans. When time spans of different durations are compared different perceptions of a physical system may result. Von Bertalanffy (1952, p. 109) states an extreme view based upon a short time span, 'In physical systems events are, in general, determined by the momentary conditions only. For example, for a falling body, it does not matter how it has arrived at its momentary position, for a chemical reaction it does not matter in what way the reacting compounds were produced. The past is, so to speak, effaced in physical systems. In contrast to this, organisms appear to be historical beings'. From this point of view, the topics of physical geology (landforms, structures, stratigraphy) are physical systems that can be studied for the information they afford during the present moment of geologic time, but they are also analogous to organisms because they are systems influenced by history. Therefore, the results obtained during short-time-span studies must be applied with care to the solution of long-time-span problems, for the history of geologic physical systems can be of prime importance for modern interpretations and predictions (Fig. 1.2).

Examples

PERIOD: An example of the importance of time as a period of record is provided by a study of long-shore drift by Kidson and Carr (1959). For the first four weeks of their study sediment movement along the beach was to the north, which was clearly opposite to the prevailing direction of sediment movement, as indicated by observed beach deposits and the configuration of the coast. After another four weeks sediment movement was to the south, which conformed to coastal conditions. If the record was for only the first four weeks of the study, it would have been difficult to explain the geomorphic and sedimentologic character of the coast. Obviously the period of record must be adequate to describe the phenomenon of concern.

Consider also Fig. 3.1 which is a record of stake exposure on Mancos Shale hillslopes near Montrose, Colorado. The stakes were driven into the slopes in order to measure erosion rates. During winter, frost heaving and swelling of the clays in the thin soil caused apparent burial of the stakes as the soil surface 'puffed up'.

37

Raindrop impact compacted the soil in early spring. It is clear from Fig. 3.1 that depending on timing and the period of record, one could conclude that erosion was rapid, measurement periods 1, 3, 5, 7, or that deposition was taking place, measurement periods 2, 6, 8, or that little was occurring, measurement period 4.

Another way of appreciating this problem is to consider intelligent organisms with different life spans. An organism with a life span of one day would likely conclude that the earth's surface was static; if the life span was 100 years, the conclusion would be that geomorphic processes are modifying the surface. An organism that lived for 10 000 years would note climatic and tectonic instability and the evolutionary development of landforms. (For a botanical example see Gleason, 1926.) Therefore, depending on the period for which data are available, extrapolation can be very meaningful or relatively useless.

Although the behavior of a landform can usually be predicted with some degree of accuracy for a decade or perhaps even a century, prediction for millenia must be based upon worst-case conditions, which involve not only geomorphic controls but also the effects of tectonic and climate change. Hence prediction is very weak for long periods, but the past does provide a guide for estimating future worst-case conditions of climate change, surface deformation and baselevel change (Schumm *et al.*, 1982). It seems that earth scientists operate at the wrong time scale for the problems that they are required to solve (Klemés, 1983; Darwin, 1956); as noted above, records are too short as are our scientific lives. Perhaps the present is too short to be a key to the past or future.

Another example of the effect of periods of record is in the interpretation of floodplains. Many floodplains form by vertical accretion of sediments (Schumm and Lichty, 1963). Each flood event deposits a layer of sediment across the valley bottom and so the floodplain is constructed vertically. However, after a period of years the lateral migration of the channel reworks the floodplain and redeposits the alluvium in a manner that indicates that the floodplain formed by lateral meander migration and lateral accretion of sediment. Therefore, in the short term the floodplain formed by vertical accumulation of alluvium, but after a period of time the evidence will be for lateral accretion.

The Hurst phenomenon (Kirkby, 1987; Church, 1980) is an important factor in evaluating a period of record, as it reflects persistence within a record. For example, wet and dry years tend to persist. Therefore, short periods tend to be less variable than long-term records. Erskine and Warner (1988) recognize alternating flood-dominant and drought-dominant periods that range in duration between 30 and 50 years. The coastal rivers of southeastern Australia respond to these climatic fluctuations by widening, among other changes, during flood-dominant periods and narrowing during drought periods. Therefore, the geometry of these rivers varies with change of flood regime and in fact, they do not become adjusted to a fixed discharge. Obviously extrapolation from short records under these circumstances is dangerous.

SPAN: In addition to the period of record, the time span considered exerts a significant influence on our perspective of a phenomenon and on interpretation of the available record. For example, for a variety of time spans and scales natural phenomenon can be classified as mega-, meso-, micro- and non-events (Table 3.2). Depending on the scale of the phenomenon, mega-events can occur during 10 million years (mountain building), 100 000 years (continental glaciation), 100 years (volcanic eruption), 10 years (meander shift and cutoff), 1 year (gully development), 1 day (landslides, rilling). A mega-event during a short period may become a non-event over longer time spans, as its effects are obliterated. For example, meander cutoff, which is such a dramatic change of river pattern, can become a non-event, and it may be undetectable after 100 000 years unless it is preserved by burial. Table 3.2 is an attempt to illustrate how an event may be a relative catastrophe during a short period of time in a limited area, but it may be insignificant for longer time spans. Note that the dimensions of the events increase with time, and the time required for their development also increases.

The dating of debris flows on Mount Shasta (Fig. 3.2) provides an example (Hupp *et al.*, 1987). The evidence for small debris flows is likely to be destroyed by medium and large debris flows. In fact, only three of the 18 small debris flows are more than 100 years old. One could conclude from Fig. 3.2 that the frequency of small debris flows increased after about 1870, which, of course, is a ridiculous conclusion, when the evidence for large debris flows extends back to 1580.

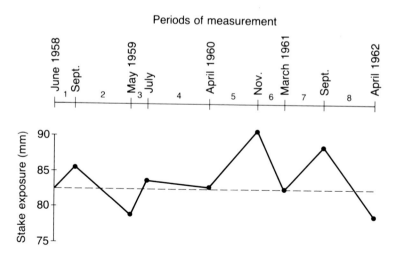

Figure 3.1. Mean stake exposure for 10 slope profiles on Mancos Shale near Montrose, Colorado (from Schumm, 1964).

Table 3.2. *Changing significance of geologic events through time (from Schumm, 1985)*

Relative magnitude of event	Time scale							
	1 day	1 year	10 years	10^2 years	10^3 years	10^5 years	10^6 years	10^8 years
Mega-event	local soil slip or flow	gully	meander cutoff	volcanic eruption	terrace formation	continental glaciation	major folding faulting	mountain building
Meso-event	rill	local soil slip or flow	gully	meander cutoff	volcanic eruption	terrace formation	continental glaciation	major folding faulting
Micro-event	sand grain movement	rill	local soil slip or flow	gully	meander cutoff	volcanic eruption	terrace formation	continental glaciation
Non-event	—	sand grain movement	rill	local soil slip or flow	gully	meander cutoff	volcanic eruption	terrace formation

Gretener (1984) refers to this type of time problem as the need to deal with the 'haze of the past', which may lead to false conclusions about rates. For example, isostatic rebound, which has occurred during the last 10000 years is still in progress in Canada and Scandinavia, and it is possible to measure it (Fig. 3.3(b)). However, this response of the earth's crust during the Holocene would appear to be instantaneous for longer time periods (Fig. 3.3(a)). Therefore, the span of time considered can significantly influence one's perception of the system that is studied. There is frequently a dichotomy in the thinking of earth scientists. On the one hand, the erosional and depositional components of the earth's surface can be viewed as evolving progressively through long periods of time, with major interruptions only as a result of climatic changes or tectonism. On the other hand, studies of erosional, transportational, and depositional processes can be made over short time spans with the assumption that landforms do not change significantly. Because of this disparate approach, the historically-oriented geologist and geomorphologist frequently find communication difficult with the process-oriented geomorphologist and engineer.

To earth scientists the passage of time or the evolutionary stage of landform development is important, but to engineers it need not be, and this leads to very different perspectives of the system of concern. For example, is stream gradient an independent or dependent variable? During a long time span, it is clearly dependent upon the discharge of water and sediment that passes through a channel. However, it is equally clear, when studying sediment transport in river channels and flumes, that the steeper the gradient of a channel or flume the higher is

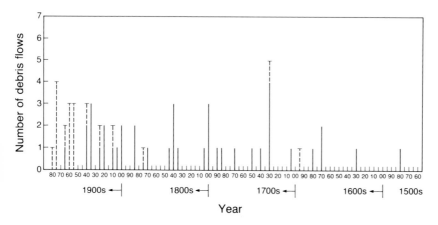

Figure 3.2. Age and frequency of debris flows on Mount Shasta dating back to 1580, as obtained from botanical evidence. Medium and large debris flows are indicated with solid lines, small flows with dashed lines. Note lack of evidence for small debris flows with increasing age (from Hupp *et al.*, 1987).

sediment transport. For example, a meander cutoff will steepen the gradient of the affected reach of channel and increase sediment transport. Obviously gradient can be both an independent and a dependent variable, depending upon the objective and the time span considered during an investigation.

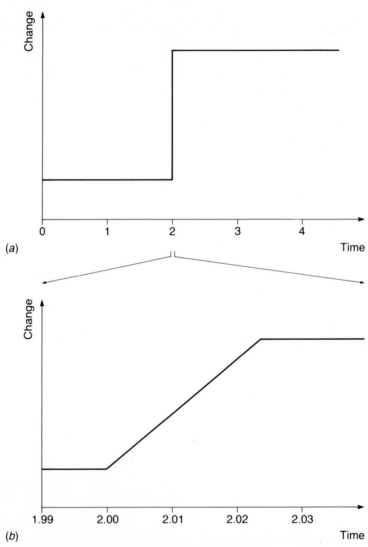

Figure 3.3. An event appears instantaneous (*a*) during a very long period but when the time scale is expanded the event appears of longer duration (*b*) (from Gretener, 1984).

42

One way to resolve this problem is to consider various equilibrium states (Fig. 3.4). If the different equilibria are assumed to represent changes of a landform through time, then they either reflect very different conditions or very different time spans. For example, if the time span is identical for each then the static and steady state equilibria (Fig. 3.4) could represent very slow changing systems, for example erosion of a drainage basin underlain by igneous or metamorphic rock, whereas the dynamic and decay equilibria could represent rapid erosion of a drainage basin underlain by shale. Conversely if the equilibria represent similar physical conditions, then the span of time involved must increase from static equilibrium to decay equilibrium. Therefore, each type of equilibrium shown in Fig. 3.4 represents spans of time of different durations (Schumm and Lichty, 1965).

Decay equilibrium represents a long span of time (geologic time), and it is

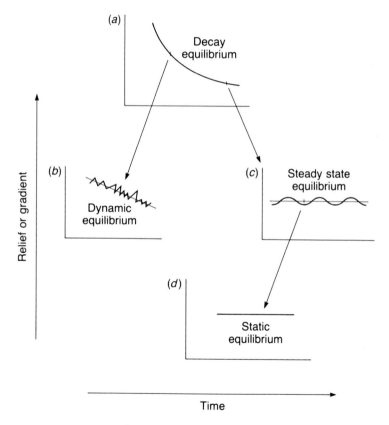

Figure 3.4. Types of equilibria: (*a*) decay equilibrium, (*b*) dynamic equilibrium, (*c*) steady state equilibrium, (*d*) static equilibrium.

analogous to the concept of a Davisian erosion cycle (King and Schumm, 1980), during which a landscape is reduced to low relief. In order to understand this concept of cyclic time consider a landscape that has been tectonically stable for a long period. Initially a certain potential energy exists in the system because of relief, and kinetic energy enters the system through the agency of climate. Over the long span of cyclic time a continual removal of material (expenditure of potential energy) occurs and the characteristics of the system change. A fluvial system when viewed from this perspective is an open system undergoing continued change, and there are no specific or constant relations between the dependent and independent variables, as they change with time (Fig. 3.4(*a*)). Dynamic equilibrium (Fig. 3.4(*b*)) represents a shorter segment of cyclic time, when the variability about a changing mean can be detected. It illustrates the complexity of landform change when viewed during an intermediate span of time, and this is the case for steady state equilibrium (Fig. 2.4(*c*)), which represents an even shorter period of cyclic time. This time division is analogous to the 'period of years' used by Mackin (1948, p. 470) in his definition of a graded stream by which he rules out seasonal and other short-term channel changes. When viewed from this perspective one sees a continual adjustment between elements of the system, for events occur in which negative feedback (self-regulation) dominates. In other words the progressive change during cyclic time is seen to be, during a shorter span of time, a series of fluctuations about or approaches to a mean (Fig. 3.4(*c*)).

Static equilibrium (present time) is very short and no change occurs (Fig. 3.4(*d*)). For example, in hydraulics steady flow occurs when none of the variables involved at a cross-section change with time.

The fact that different time spans are considered by different specialists opens the possibility of serious misunderstanding and lack of communication which can be a serious impediment in planning. For example, an engineer who consistently deals with static and steady-state equilibria may have considerable difficulty in understanding the geologist who deals more often with the cyclic or graded time spans. The difference was expressed by Von Bertalanffy (1952) in the quotation that was cited above (p. 37). Is it only the mechanics of a physical system that is under consideration, or does the physical system have a history? This is very important because cause and effect relations can change depending upon the time span considered.

This is best demonstrated by comparing the conflicting conclusions that could result from studying fluvial processes in the hydraulic laboratory and in a natural stream. The measured quantity of sediment transported in a flume is dependent on the velocity and depth of the flowing water and on flume shape and slope. An increase in sediment transport will result from an increase in the slope of the flume or an increase in discharge. However, in a natural stream during longer periods of time, it is apparent that mean water and sediment discharge are the independent variables, which determine the morphologic characteristics of the reach of stream under consideration and, therefore, its flow characteristics. Furthermore, over very long periods of time (cyclic) the independent variables of geology, relief, and climate determine the discharge of water and sediment with all other morphologic and hydraulic variables being dependent. Both Mackin (1963) and Kennedy and Brooks (1965) used this identical example to illustrate the need to consider how time spans are relevant to the explanation of fluvial phenomena. Kennedy and Brooks (1965) state it thusly (Q and Q_s are water and sediment discharge):

> The time scale under consideration is also important in classifying Q and Q_s as dependent or independent variables. Streams are seldom if ever in a steady state (because of finite time required to change bed forms and depth) and transitory adjustments are accomplished by storage of water and sediment. Water storage is relatively short (hours and days) and occurs simply by the increasing of river stage or over-bank flooding; sediment storage ($+$ or $-$) occurs by deposition or scour. Thus for the time term, Q_s may be considered a dependent variable, with departures of the sediment inflow from the equilibrium transport rate being absorbed in temporary storage (for months or years). But in the long term the river must assume a profile and other characteristics for which on the average, the inflow of water and sediment equals the outflow; consequently for this case (called a 'graded' stream by Mackin), Q and Q_s are . . . independent variables.

Depending on one's viewpoint, the earth's surface is always in one stage in a cycle of erosion or it is in dynamic equilibrium with the forces operative. These views are not mutually exclusive. It is just that the more specific we become the shorter is the time span with which we deal.

Earth scientists are not alone in recognizing that consideration of different spans of time change the perspective of the investigator (Clark, 1987; van der Heijde, 1988). This is true also in zoology (Udvardy, 1981), climatology (Mitchell, 1976; Webb *et al.*, 1985), and botany (Delcourt *et al.*, 1983).

Thornes and Brunsden (1977) appreciate the impact of time spans on development of models of landform change. They consider short-term processes studies as 'floating-equilibrium models'. That is, the understanding of an erosional or depositional process can be extrapolated to any period of geologic time. This is true uniformity.

Over a longer period of time progressive change can be observed (dynamic equilibrium), and a 'relative-time model' of this change can be developed that also can be used to postdict and predict. These relative time models, which incorporate the floating equilibrium models, when applied to a specific area, permit development of the erosional history of that area, for a specific period of time. 'Fixed-time models' therefore, are unique to one area.

Another way of viewing the significance of time spans is to consider the stratigraphic record. During a short time span, which is represented by a small part of a

45

formation, the record can show continuous deposition. During a longer time span, the sedimentation rates will fluctuate, and evidence for changing rates of deposition will be evident. During very long spans, the record will be punctuated by episodes of non-deposition or changes of sediment character. Depending on the time span considered, the depositional model will vary from continuous deposition to episodic deposition.

The obvious result of this is that, as time spans increase, average rates of change decrease. For example, denudation rates (Gardner *et al.*, 1987), sediment accumulation rates (McShea and Raup, 1986; Sadler, 1981; Plotnick, 1986), and evolutionary rates (Gingerich, 1976) all decrease with increasing measurement interval or time span (Gardner *et al.*, 1987). This of course, leads to discussions of the completeness of the geologic record (Ager, 1981; McShea and Raup, 1986).

It is also important to think of different spans of time for practical purposes. For example, most landforms, when perturbed, will follow an evolutionary development returning to an approximate initial condition (negative feedback), or they will find a new condition of stability. The example of incised channels (gullies, arroyos, etc.), that is discussed in the following paragraph, shows how viewing the change in terms of cyclic time will aid in controlling the unstable condition or at least in recognizing when the expenditure of funds will be most beneficial.

Time is an important variable in the development of an incised channel and, therefore, it should be an important variable in any scheme to curtail gully erosion and to reduce sediment loads. Figure 3.5 is a conceptual diagram that shows the change in sediment yield and incised channel (gully) drainage density (length of gullies per unit area) with time. In a drainage basin that has been rejuvenated and in which gullies are developing, sediment production will increase, as the length of incising channels increases (Fig. 3.5, times 1 to 4). However, at time 4 maximum headward growth of the channels has occurred, and they begin to stabilize between times 4 and 7, when there is an increase in the length of relatively stable reaches, and the length of active reaches decreases. By understanding this cycle of channel incision and gullying from initial stability (time 1) to renewed stability (time 8), it is possible to select spans of time in the cycle when land management and incised channel control practices will be most effective. For example, gullies just initiating (times 1 or 2) and gullies almost stabilized (times 6, 7 or 8) will be the most easily controlled by artificial means. The efforts at times 1 and 2 will be most effective in preventing erosion, whereas efforts at times 6 and 8 will have little effect, as the channels are stabilizing naturally. At time 4 control will be difficult and expensive. Obviously, consideration of this complex evolving system for only short periods of record and short time spans can yield erroneous conclusions, and this will be demonstrated at a regional scale in Chapter 4.

In conclusion, both the period of record, as well as the time span under consideration, can be critical to the understanding of natural phenomena.

Space

Definition
Space is the three dimensional field in which natural phenomena function and occur, and in which the subject of an investigation exists. There are two aspects of the space problem, scale and size. Scale involves the resolution at which an object is viewed, and size involves simply the comparison of large and small things.

Statement of problem
At a small or coarse scale details are not observed, but at a large or fine scale much more of an object can be seen. The complexity of the subject will increase as size increases (small to large) and as scale becomes larger (low to high resolution). Therefore, it may not be possible to extrapolate confidently from small to large or vice versa, and conclusions reached at poor resolution should be held with little confidence.

Examples

SCALE: There is no obviously preferred scale that can be used for the study of a natural system. However, we have the best understanding of things that are within the 'human-scale', phenomena which are accessible to humans directly

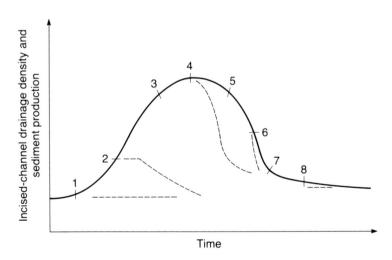

Figure 3.5. Hypothetical change of sediment production and incised channel (gully) drainage density with time. Dashed lines indicate effect of gully-control structures at various times during channel evolution.

through their unaided senses. These are roughly from ¹⁄₁₀ of a millimeter to a few kilometers in space and from ¹⁄₁₀ of a second to a few decades in time (Klemés, 1983). For the processes at this scale we have an intuitive feel.

The scale used should depend upon the problem under consideration, which will determine the component of the system or the size of the unit studied. Therefore, many scales of size and time can be used depending upon what aspect of a system is being considered. For example, the fluvial system can be considered at different scales and in greater or lesser detail depending upon the objective of the observer. A large segment, the dendritic pattern, is a component of obvious interest to the geologist and geomorphologist (Fig. 3.6(*a*)). At a finer scale is the river reach of Fig. 3.6(*b*), which is of interest to those who are concerned with what the channel pattern reveals about river history and behavior and to engineers who are charged with maintaining navigation and preventing channel erosion. A single meander can be the dominant feature of interest (Fig. 3.6(*c*)), when studied by geomorphologists and hydraulic engineers for information that it provides on flow hydraulics, sediment transport, and rate of bend shift. Within the channel is a sand bar (Fig. 3.6(*c*)), the composition of which is of concern to the sedimentologist, as are the bed forms (ripples and dunes) on the surface of the bar (Fig. 3.6(*d*)) and the details of their sedimentary structure (Fig. 3.6(*e*)). These, of course, are composed of the individual sediment grains (Fig. 3.6(*f*)) which can provide information on sediment sources, sediment loads and the feasibility of mining the sediment for construction purposes.

As the above example demonstrates, the components of the fluvial system can be investigated at many scales, but no component can be totally isolated because there is an interaction of hydrology, hydraulics, geology and geomorphology at all scales, which emphasizes that the entire fluvial system must be considered, although only a small part of it may be under investigation. The conclusion must be that it is important to consider a problem at all available spatial scales.

SIZE: As size increases the complexity of a feature increases. A small drainage basin can lie within one climatic region, form on one lithologic unit and be subjected to one type of land use. Large drainage basins can span climatic, lithologic, and land–use boundaries, and thus they are more complex. Size alone may not be a problem, because in very simple situations small size can yield valuable information about large size, but as complexity increases extrapolation from small to large may be of little value.

As size changes explanations may change (Arnett, 1979; Phillips, 1988). For example, although local variations of stream gradient can be explained by variations of bed–material size, the gradient of a longer river segment is better explained by water discharge (Penning-Roswell and Townshend, 1978). Morgan (1973) demonstrates the relative importance of size on drainage density in Malaysia. In small second–order drainage basins both climate and lithology

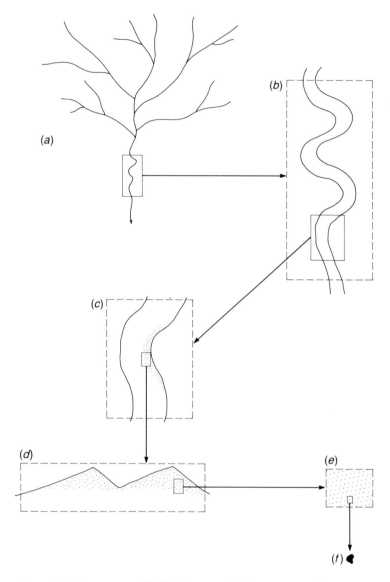

Figure 3.6. Components of the fluvial system: (*a*) drainage network; (*b*) meandering river; (*c*) meander and point bar; (*d*) bed forms; (*e*) cross stratification; (*f*) sediment grain (from Schumm, 1985).

determine drainage density, but lithology dominates at the meso-scale (Klang River basin) and climate at the macro-scale (west Malaysia).

Komar (1983) provided an example of the difficulty of extrapolating from small to large by examining rhythmic shoreline features that range in size from a few centimeters to several kilometers. Although all of the features have similar morphology, they are formed at different scales by different processes (see problem of convergence). The smallest of these features are beach cusps (Fig. 3.7(*a*)) that have spacings less than 25 m. They are formed by standing waves that appear as a set of waves with their crests normal to the shoreline (edge waves). Beach cusps, therefore, are the result of hydraulic wave action.

Cuspate features with a larger than 25 m spacing can be formed by rip-current action (Fig. 3.7(*b*)). Cuspate features with spacings of between 100 to 2000 m can be related to crescentic bars that modify wave refraction and diffraction patterns,

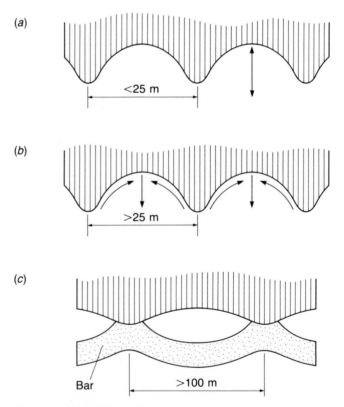

Figure 3.7. Rhythmic shoreline forms of decreasing size: (*a*) beach cusps (< 25 m), (*b*) rip-current cusps (> 25 m), (*c*) crescentic-bar cusps (100–2000 m) (from Komar, 1983).

which, in turn, alter the shoreline configuration (Fig. 3.7(*c*)). Depending on energy conditions the above processes can produce features from small to large, and therefore, any classification based upon size alone 'must fail and simply lead to confusion' (Komar 1983, p. 106).

It is well known that as drainage basins increase in size the sediment delivered from each unit of drainage area will usually decrease (Fig. 3.8). This is because sediment production is greatest in the steeper upper parts of a drainage basin. In addition, gradients decrease and valleys widen, and the opportunity for sediment storage increases downstream. Therefore, as drainage basin size increases, the unit sediment production and sediment yield decrease.

On the other hand, if a drainage basin has been rejuvenated by baselevel change or uplift this relation could, in fact, be reversed, and large rivers in British Columbia, that are incising into and reworking sediment that was delivered to the valleys by glaciers during the Pleistocene epoch, do indeed show an increase of sediment per unit area with increased drainage area (Church and Slaymaker, 1989).

Summerfield (1982) points out that the direction of research can also be dominated by the size of the units investigated. Recent geomorphic investigations have concentrated on the erosional and depositional processes that modify components of the landscape (hillslopes, channels, etc.). Therefore, origin of relief (Table 3.2) and tectonic processes are ignored in these studies. However, as the subject of investigation moves from specific landforms to larger landscape units the influence of specific erosion processes becomes less significant. On the scale of a mountain

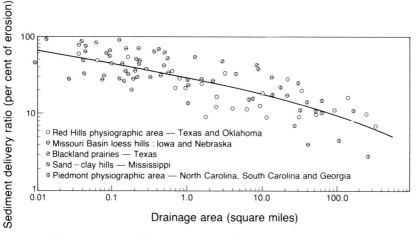

Figure 3.8. Effect of drainage area on sediment delivery ratio (ratio of sediment yield from a drainage basin to sediment production within the basin) (from Boyce, 1975).

51

range the surface processes play a secondary, though important role, as tectonics becomes dominant (see Table 3.2).

If there appears to be a similarity in problems of scale and problems of time this is because they cannot really be separated (Table 3.2). For example, Trudgill (1976) argues that the rate of change of landforms of different size varies with time as climates change. He cites work in a limestone area of Iran where the rounded hills and karst features are relict from a wetter climate, when solution was the dominant process (Waltham and Ede, 1973). However, small solutional forms are now being destroyed by frost action. He concludes that the present climate is 'not expressed in the larger-scale forms but only in the small-scale forms'. Figure 3.9 illustrates Trudgill's concept of micro, meso and macro landform response to climate change. Obviously weaker materials and sensitive landforms will respond much more quickly to a change than will landforms composed of resistant materials.

As the size and age of a landform increases, fewer of its properties can be explained by present conditions and more must be inferred about the past. Figure 3.10 is an attempt to suggest the range of historical information that is needed to explain specific phenomena. Thus microfeatures and events such as sediment movement and bed forms in a river are understandable in the light of recent experience without historical information. However, channel morphology may have a sizeable historical component. For example, rivers flowing on valley or alluvial plain surfaces, the slopes of which were determined by Pleistocene events,

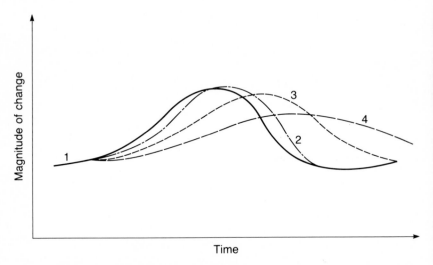

Figure 3.9. Lag of the response of landforms of different size to climate change. Curve 1 represents climate change, curve 2 microforms, curve 3 mesoforms, and curve 4 macroforms (after Trudgill, 1976).

are significantly influenced by history (Schumm, 1977). Large features such as drainage networks that are structurally controlled (trellis, rectangular), and mountain ranges will obviously be explained predominantly by historical information.

The longer the time span and the larger the area, the less accurate will be predictions and postdictions that are based upon the present. Therefore, extrapolation of modern records involves risk, and geomorphic predictions for periods in excess of perhaps 1000 years should be based upon worst-case conditions of climate change, tectonic activity and baselevel change. This, of course, involved an understanding of the past. The past reveals much about major features and slowly acting processes but little about mesofeatures and nothing about microfeatures, which become non-events as a result of rapid erosion or deposition (Table 3.2).

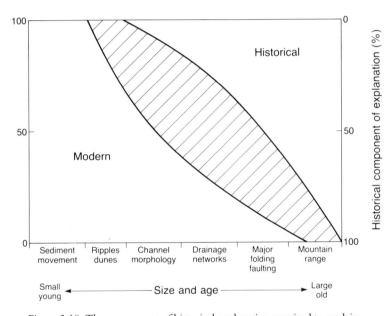

Figure 3.10. The component of historical explanation required to explain events of increasing magnitude and age. Upper curve shows maximum modern component. Lower curve shows maximum historical component. Zone between curves is range of variability. For example, depending on circumstances, channel morphology may have from about 10 per cent to 60 per cent historical explanation. Modern bed forms in rivers can be explained totally by hydraulics (100 per cent) or partly by the nature of sediment derived from older deposits (25 per cent (?) historic). Mountain ranges may be explained totally in terms of past orogeny and denudation (100 per cent historic) or partly in terms of plate movements, neotectonics, and modern erosion rates (50 per cent (?) modern) (from Schumm, 1985).

The present reveals information on micro- and mesofeatures, but perhaps little about macrofeatures.

An additional topic that should be considered here is sample size, which is analogous to the period of record problem that was described earlier. Physicists and chemists deal with large amounts of data from which good statistical predictions can be made. Their experiments are also readily repeatable. In contrast, the earth scientist may have a very small sample, and therefore, the probability that a prediction is accurate is not strong. Also the prediction is usually for a single feature rather than for a very large sample. For example, it is not possible to predict which individual will die of lung cancer, although it is possible to predict that the probability of a heavy smoker contracting lung cancer is high. The Tobacco Institute argues that it has never been proved that any individual died from smoking. They are right, but the probability that the smoker died of cancer from smoking is high.

In conclusion, scale and size are very important problems in any attempt to extrapolate. Extrapolation and explanation in space is as difficult as extrapolation and explanation for different periods and spans of time.

Location

Definition
Location is a site or place at which observations are made in a natural system. This problem may seem similar to that of size, but it really involves extrapolating from one location to another, for example, from one tributary to another in a single drainage system (Fig. 3.6(*a*)). There are two aspects to this problem, the actual difference between different places and the different perspectives that are obtained by workers who were trained and operate in different geographical locations.

Statement of problem
Even the smaller components of a landscape such as first-order stream may have a considerable range of potential energy from mouth to drainage divide. The morphology varies and the materials and energy flow vary from place to place (Graf, 1982). As noted before, sediment delivery ratios are usually below 1.0 (Fig. 3.8), which indicates sediment storage in the drainage basin. In other words, some portions of the drainage network are eroding whereas others are aggrading. Therefore, to assume that what is observed or measured at one location in a geologic system occurs elsewhere may be in error.

Examples

PLACE: The question of whether we can extrapolate from one place to another is a simple problem in comparison to the problems of time and space. The answer depends on how similar the two locations are. In high-energy landscapes

54

not all components of a drainage network are functioning in phase. Some tribu-taries may be stable, others aggrading, others degrading, depending on local cir-cumstances or on the rate at which they respond to changes in the main channel (Schumm and Hadley, 1957; Schumm, 1977; Waters, 1985). Over a long period of time the system will appear to be in equilibrium, but the details of alluvial stratigraphy and the terrace sequence will show many variations and many anom-alies (McDowell, 1983), which occur because of energy differences within the basins and the random occurrence of hydologic and meteorological events. Therefore, conclusions may depend on what part of the system is being studied. For example, rejuvenation of a drainage basin by baselevel change will cause a wave of accelerated erosion to advance headward through the basin. Depending on the size of the basin, the lag time for features near the drainage divide may be very long, and events in one part of the basin may be very different from those occurring elsewhere for hundreds or even thousands of years.

An exceptional example is provided by the studies of Bergstrom (Bergstrom and Schumm, 1981) in the badlands of southeastern Wyoming (Fig. 3.11). In the headwater channels of the small badland drainage basin (zone 1) sediment is stored during winter and early spring (periods 1 and 2). Downstream in the higher-order channels (zones 2 and 3) small amounts of runoff cause scour and channel incision. During the heavy rains of late spring and early summer the sediment in zone 1 is removed by channel incision (periods 3 and 4), and this sediment is deposited in zone 2 (period 3). Subsequent removal of the stored sediment in zone 2 (period 4) causes aggradation downstream in the main channel (zone 3, periods 3 and 4). The processes acting in the components of this small badland drainage network are markedly out-of-phase.

Further examples of the effect of location relate to Pleistocene climate change. Studies in midcontinent USA and Europe indicated that a colder wetter climate was typical, but, it is now known that cooler, drier climates were common else-where, and perhaps cooler climates with no change in precipitation also occurred (Lamb, 1977). Therefore, attempts to extrapolate landform history and especially channel behavior from even one location to another may lead to error. For example, correlation or prediction for polyzonal rivers, for example a river that flows from glacial regions through humid and semi-arid regions to the sea, is dif-ficult. Through time, the lower part of the channel will be dominated by sea-level fluctuations, the upper part by glacial events, and the middle part by climatic fluctuations from semi-arid to sub-humid or arid. The correlation of terraces in such a situation obviously is a problem, and attempts to correlate Pleistocene terraces between the upper and lower Mississippi River failed (Starkel, 1979).

Knox (1983) in an attempt to correlate alluvial deposits of late Pleistocene and Holocene age in the US found that there were general relations, but correlation was difficult. However, his studies revealed that rivers in the humid eastern US appear to respond to climate and hydrologic change by lateral channel shift.

Whereas in the semiarid western US the adjustment was more likely to involve vertical change by aggradation or degradation.

Rivers usually increase in size in a downstream direction, as a larger channel is required to convey the increasing discharge. In general, channel width and meander dimensions increase as about the 0.5 power of discharge, and channel depth increases as about the 0.4 power of discharge. However, when river discharge decreases in a downstream direction, channel dimensions also decrease. For example, the Finke River in central Australia flows from its source to the McDonald Range as a wide sandy channel, but the channel becomes smaller as the river flows into the Simpson Desert, and it eventually disappears. The Sacramento River also decreases in size downstream because flood waters are diverted naturally into adjacent flood basins and floodways. Thus the simplest assumption may be affected as size and location change.

Location is also very important in stratigraphic and sedimentologic studies. That is, conclusions reached at one location on a delta or alluvial fan will probably be different if another location were studied.

PERSPECTIVE: Another important aspect of location is the physical location of an investigator on the globe and the investigator's experience or lack of it from other locations. Some of the great geological controversies have been the result of the dominance of a location on the thinking of an individual. An excellent example is provided by Hallam's (1983) review of the controversy between the Neptunists and Plutonists in the eighteenth century. Werner worked in Saxony in central Europe, where horizontal or subhorizontal layers of basalt capped some hills and were also interbedded in sedimentary strata. Desmarest worked in the Auvergne area of the Massif Central of France, where he was able to trace basalt flows back to volcanic craters. Werner thought that basalt was a sedimentary rock, but Desmarest proved it was volcanic. According to Hallam (1983, p. 9) Desmarest admitted that if confronted only with the evidence of the Saxon Hills he could not have determined that basalt was volcanic.

The explanation of glacial till as the result of glacial action rather than as the result of great floods came from Switzerland, a country with modern active glaciers. Hallam (1983, p. 67) suggests that 'if such a keen and perceptive observer as Hutton had spent some time in Switzerland the glacial theory might have been proposed much earlier than it was', and indeed, the drift theory was abandoned sooner in Switzerland, where the evidence against flooding and for glacial action was apparent.

Earth science investigations are geographically determined (Westgate, 1940). Rocks vary greatly from place to place, in age, in structure and in surface expression; and in most cases rocks and other natural phenomena have to be studied where they are found. As an example, stratigraphy began in central England under William Smith and with Cuvier and Broignart in the Paris basin, where gently inclined beds crop out so that the order of vertical succession is easily seen. Major contributions to pre-Cambrian geology were made by the geologists of Canada, the northern United States and Scandinavia, working in these great pre-Cambrian areas. Glacial geology, as might be expected, has inevitably had its beginnings in a region of existing glaciers, the Alps. Of course, glaciers exist in other parts of the world, in Alaska, Chile, India, Greenland, Antarctica, but these regions were practically unknown, and far removed from any center of scientific activity in the nineteenth century. The Alps were near at hand and accessible. If there was to be a father of glacial geology, the odds were heavily in favor of some scientist living near their borders; he might have been German, Austrian, French or Swiss; he was the Swiss, Agassiz.

The early support for continental drift came largely from those with field experience in the southern hemisphere because it is there that the evidence for it

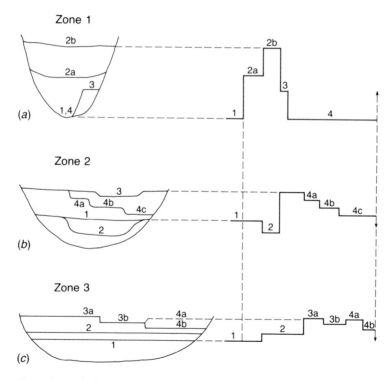

Figure 3.11. Idealized cross-sections showing changes in zones, 1, 2, 3 during one season in the Kraft Badlands. The diagram to the right of each cross-section illustrates channel behavior in each zone. Numbers represent seasons from late winter (1) to early spring (2), late spring (3) and early summer (4) (from Bergstrom and Schumm, 1981).

was strongest (Giere, 1988, p. 240). The author remembers vividly his conversion, when he rubbed the glacial striations beneath the Dwyka Tillite near Durban, South Africa and was forced to conclude that the glacier had advanced from what is now the Indian Ocean with a land mass no closer than Australia.

A more recent problem with regard to both location and perspective relates to bank stability of the Ohio River. Twenty-two landowners claimed that the erosion of their property was caused by the raising of water levels behind navigation locks and dams. In order to maintain navigation on the Ohio River during low water a series of low dams with locks maintain a minimum navigation depth of nine feet. The pool level behind the dam, therefore, never falls to the old low-water levels. It was alleged that the maintenance of the pools at a constant level causes bank erosion by wave action. Preliminary studies showed that, indeed, erosion was occurring on the litigants' lands, and their claims seemed valid. However, when the river as a whole was considered, rather than just twenty-two limited portions of the bank, it became clear that the river has eroding, stable and healing banks, and the type and extent of erosion could be expected. In fact, much of the erosion was due to the landowners' activities behind the bankline, which added water to the banks and caused slumping well above the pool level. In this case the ability to consider the entire river rather than a few specific locations permitted the development of a strong argument that the bank erosion was natural or that in some cases it was induced by the landowners themselves. The landowners lost the case because the judge found that the geomorphic arguments were convincing, but the landowners probably were not convinced because of their limited perspective of the system. Woodall (1985) has referred to this in mineral exploration as 'limited vision'.

In conclusion, predictions based upon data from one location may not be valid elsewhere. Extrapolation to other locations is as subject to error as extrapolation in time and space. Furthermore, an investigator's experience and perspective may be crucial in solving a problem, or, indeed, the investigator's bias may prevent a solution.

PROBLEMS OF CAUSE AND PROCESS

Convergence

Definition
Convergence refers to a situation when different processes and different causes produce similar effects. This is often referred to as equifinality (Chorley, 1962).

Statement of problem
If different causes or processes produce similar effects then the use of analogy breaks down. Therefore, it is sometimes difficult to infer processes and cause from effect (Pitty, 1982, p. 44; Chorley, 1962, p. 138; Chorley and Kennedy, 1971,

58

p. 294), and attempts to do so have been termed the genetic fallacy (Harvey, 1969, p. 80, p. 409).

Examples

CAUSE: A good example of convergence due to different causes is channel incision and terrace formation, which can be due to baselevel lowering, climate change, tectonics, and in modern times changes of land use. At least four different causes produce the same effect, channel incision.

This type of convergence problem became very obvious during the early days of planetary studies, when the origin of sinuous rilles on the moon were attributed to flowing water (Fig. 3.12). However, a variety of volcanic features was found to resemble fluvial landforms, and lava channels, collapsed lava tubes and fluidization features were determined to be the cause of the lunar channels (Schultz, 1976).

Braided stream patterns can also be caused by different controls. For example, braided streams result from aggradation, but they also can be 'stable' with the braided morphology being the effect of high bedload transport on steep gradients. Under similar conditions, flashy discharges may maintain a braided channel, when elsewhere with more uniform hydrologic conditions the channel meanders.

On a global scale, Mörner (1984) points out that ocean-level changes (eustasy) can be glacial-eustatic, as a result of climate change and continental glaciation, tectono-eustatic, as a result of changes in ocean-basin volume by tectonics, and geoidal-eustatic, as a result of gravitational and rotational changes of the geoid.

PROCESS: Different processes can also produce similar appearing landforms. Gently sloping piedmont surfaces (pediments, alluvial plains) can be formed by parallel slope retreat, by lateral planation by streams and by other more complicated processes (Schumm, 1977, p. 286) as well as by deposition.

Dendritic drainage patterns develop on relatively homogeneous rocks without structural control, but how meaningful is this definition when dendritic fracture patterns also occur (Fig. 3.13)? The question arises how many dendritic drainage patterns are controlled by dendritic fracture patterns rather than fluvial processes? The answer is probably few, but the possibility exists.

Scarps are formed when a resistant caprock protects underlying weaker rocks and provides a talus apron at the scarp base. The process of retreat is by mass movement, as rocks fall from the caprock. However, in the Colorado Plateau similar appearing scarp forms develop without talus, and the local slope is formed on bedrock by rainwash, creep and rilling (Fig. 3.14). In fact, it can be difficult to determine from the appearance of a landform if it is depositional or erosional, for example, alluvial fans, rock fans or pediments.

Folding and faulting, which are usually the result of tectonic activity, can also

59

be developed by mass movement processes. For example, ridge-top graben-like features (Fig. 3.15(*a*)) can be formed by lateral spreading away from a topographic high (sackungen), and a series of gentle anticlines and synclines can be formed by cambering, which is the draping of competent sedimentary units over hillsides, when underlying easily deformed sediments (clay) creep toward the valley (Fig. 3.15(*b*)). The synclines always occupy the valleys and anticlinal axes follow divides (Judson, 1947). The word cambering is from the architectural term camber, which is defined as the condition of being slightly arched. Careful field work would be required in order to discriminate these mass movement produced 'structures' from the real thing.

Terraces also can be formed in different ways, and it may be difficult in some

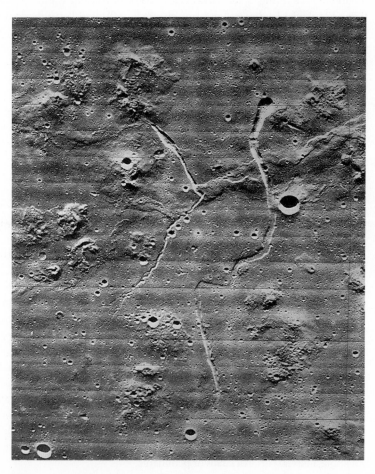

Figure 3.12. Sinuous rilles on Moon (NASA, Lunar Orbiter V, V-51, M213).

instances to determine if a terrace is formed by river action or by the stripping back of weaker rocks above a more resistant layer. The problem would be greatly intensified if the resistant layer were a fluvial conglomerate (Fig. 3.15(c)). Casual inspection might suggest fluvial terraces, but a stratigraphic study would show that the conglomerate continued into the hillside and the terrace is a stripped surface (J. Loen, 1988, personal communication).

Haines-Young and Petch (1983) argue that the concept of equifinality is invalid as a research tool because careful study of a situation may show that the similarity between landforms is more apparent than real. For example, Figs. 3.13, 3.14 and 3.15 illustrate situations where similarities are only apparent, and for at least Fig. 3.14 even a casual field check would reveal the differences. Nevertheless, often this is not done, which is the reason that convergence should be included as a problem always to be considered.

Haines-Young and Petch (1983) also state that, although Meteor Crater and Gilbert's study of it apparently provide an example of convergence, the significance of convergence would disappear if the craters were given specific names, for example, volcanic crater, meteor crater, phreatic crater. However, the point is that the investigator needs to know that craters can be formed by different processes and then, as done by Gilbert, testing of each mode of formation is undertaken.

In conclusion, the similar results from different causes and processes make interpretation of earth surface features difficult, and therefore, a fragmentary record from the geologic past or limited observations at the present may be an

Figure 3.13. Dendritic fractures in lake ice.

inadequate base upon which to postdict or predict. Nevertheless, this adds to the excitement of earth science as illustrated by the following quotation (Hartshorn, 1967, p. 46) from a letter from T. C. Chamberlin to N. S. Shaler in 1885. 'I esteem it an especial intellectual beneficence that nature is sufficiently rich in method to accomplish very similar ends by diverse means, and very diverse ends by similar means'. Obviously under these conditions extrapolation must rest upon a careful study of the system of interest.

Divergence

Definition
Divergence is the opposite of convergence and refers to a situation when similar causes and processes produce different effects.

Figure 3.14. Scarp with basal bedrock slope. A cursory inspection would suggest that the basal slope is a talus accumulation.

Problems of cause and process

Statement of problem

If similar processes or causes produce different effects then homology breaks down, and it is difficult to determine with confidence the cause of a phenomenon. For example, a climate change may trigger massive landslides in one area, gullying in another and a limited response elsewhere. Therefore, existing conditions must be thoroughly understood before extrapolation can be attempted.

Examples

CAUSE: An unusual example of different effects produced by the same cause is when baselevel lowering or increased stream discharge causes aggradation. The usual result, of course, is degradation. In this case main channel incision causes tributary incision, which delivers a larger and perhaps coarser sediment load to the main channel, which in turn aggrades. Hence, in the same channel the same cause produces both erosion and deposition.

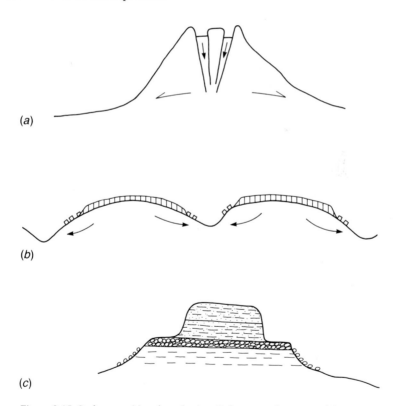

(a)

(b)

(c)

Figure 3.15. Sackungen (*a*) and cambering (*b*) form pseudostructural features. A fluvial conglomerate forms pseudofluvial terraces (*c*).

A global example is the effect of glacio-eustasy on sea-level change (Clark *et al.*, 1978). With melting of the Pleistocene continental ice sheets the assumption is that a global sea-level rise will submerge all coastlines. However, the results are quite variable (Fig. 3.16). As a result of continuous isostatic uplift following melting of the continental ice sheets, there are raised shorelines in zone I. As a result of a collapsing forebulge in front of the ice sheets, there are submerged shorelines in zone II. In zones III, IV and V relative sea-level movements reflect differential effects of glacial and isostatic response. For example, zone III contains an emerged

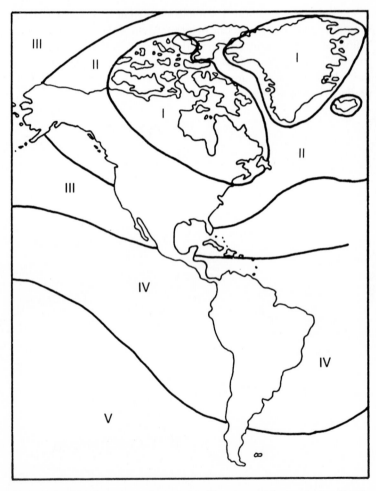

Figure 3.16. The five zones, where sea level has changed during the last 16000 years. Emerged coasts occur in zones I, III, and V, submerged coasts are characteristic of zones II and IV (from Clark *et al.*, 1978).

beach, which formed a few thousand years ago, whereas in zone IV there has been continual submergence, and zone V shows evidence of submergence that was followed by minor emergence after meltwater ceased to flow into the ocean.

Hydrologic and geomorphic studies show that both sediment yield and drainage density (length of channels per unit area) under natural conditions are a maximum in semiarid areas (Langbein and Schumm, 1958; Gregory and Gardiner, 1975). Therefore, a climate change that results in greater precipitation everywhere will have very different effects in arid, semiarid and humid regions (Fig. 3.17). For example, an increase in precipitation in arid regions will significantly increase drainage density and the export of sediment from that area. A similar change in amount of precipitation in semiarid regions will produce less sediment, due to the increased vegetative cover, and in humid regions drainage density will decrease because greater vegetation density and better-developed soils will reduce the number of small channels. It should be noted that the sediment yield curve is based upon data from continental United States from drainage basins with natural vegetation. That is, human influences are minimal.

Another example is provided by river channel patterns, when an increase of

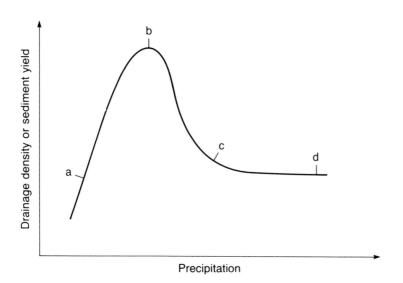

Figure 3.17. Diagram showing how drainage density (total length of stream channels divided by drainage area) and sediment yield vary with mean annual precipitation under natural conditions. With an increase of precipitation, drainage density and sediment yield increase to a maximum (a to b) in semi-arid regions, decrease in subhumid regions (b to c) and then remain relatively constant (c to d) in humid regions, with other variables remaining the same.

energy, due to increased discharge or perhaps to active tectonic steepening of the valley floor, causes a straight stream to begin to develop a sinuous pattern, a mildly meandering stream to become more sinuous, a highly sinuous stream to become braided, or a braided stream to remain braided (Fig. 3.18).

PROCESS: All of the major erosional transportational and depositional processes, aeolian, fluvial, glacial and marine produce different effects depending upon the conditions under which they operate. For example, a great variety of dune forms depend on availability of sand and the prevailing wind direction. A continuum of river types from low to high sinuosity and from braided to anastomosing all form by fluvial processes but under different discharge and sediment load conditions. This aspect of divergence is well known and in most cases the differences in dunes, rivers, glacial landforms and shorelines can be explained and is expected. Therefore, when considering divergence, process is less of a problem than cause.

In conclusion, the different effects produced by similar causes and processes can significantly complicate landform interpretation and the geologic record and can make extrapolation difficult.

Efficiency

Definition
Efficiency refers to the ratio of the work done to the energy expended. Thus, it refers to the impact of an event or series of events on a system. This is in contrast to effectiveness, which is the degree to which the event or energy expenditure had the desired response. Obviously in natural systems there is no 'desired' response.

Statement of problem
It is usually assumed that the more energy expended the greater the response or the greater is the work done. However, when more than one variable is acting or when a change of the independent variable, such as precipitation, has two different effects, for example, increased runoff and increased vegetation density, there may be a peak of efficiency at an intermediate condition. This complicates the development of cause and effect relations.

Examples
Wolman and Miller's (1960) magnitude–frequency concept is an example of how the 'relative amount of "work" done during different events is not necessarily synonymous with the relative importance of these events in forming a landscape'. They use sediment transport as an example of work done, and they show that, although large floods carry tremendous sediment loads, they are infrequent, and over a long period of time the most sediment movement will be accomplished

during events of moderate magnitude and frequency (Fig. 3.19(*a*)). However, if the threshold of sediment movement is high because the sediment is cobbles or boulders instead of sand (Fig. 3.19(*b*)) maximum sediment movement can be associated with large magnitude events. Of course, a large precipitation or flood event can cause considerable geomorphic change that will be healed by subsequent smaller events, but large events are not always efficient.

Beven (1981) points out that the effectiveness of floods of a given magnitude will vary significantly depending on preceding events, or 'event ordering'. That is, effectiveness or efficiency of an event may be greater if it is preceded by a large event rather than a small one, but even a small event will be more efficient if it follows a preceding event quickly. Antecedent precipitation is recognized as a regulator of runoff from a later storm. Thus, the efficiency of flood events of a given magnitude can vary greatly with time. In conclusion, it cannot be assumed that a steady increase of a controlling variable will produce a corresponding increase in a dependent variable.

Suzuki and Takahashi (1972, 1981) carried out rock abrasion experiments in which an air jet with entrained sand grains was directed toward polished rock surfaces. They were able to show that the rate of abrasion of the rock surface (V) increased with the rate of sand feed (Q), but abrasion efficiency ($m = V/Q$) was a maximum at relatively low rates of sand feed (Fig. 3.20). The decrease in efficiency

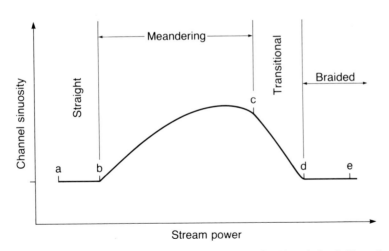

Figure 3.18. Diagram showing how sinuosity (channel length divided by valley length) varies with stream power (tractive force times velocity of flow). With an increase of stream power or velocity sinuosity remains constant at low values (a to b), increases with meandering (b to c), decreases through a transition from meandering to braided (c to d) and then remains braided (d to e) (from Schumm and Khan, 1971).

67

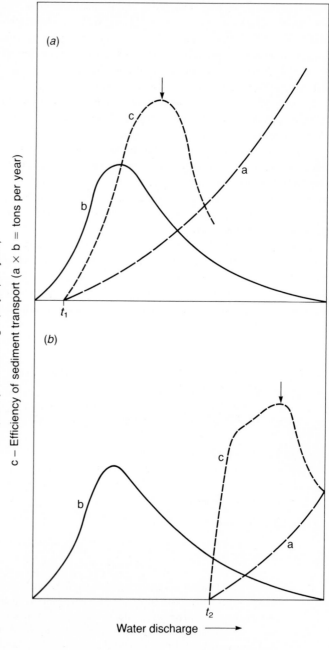

is probably due to rebound effects, as particles rebound and interfere with incoming particles.

Abrasion by wind shows a range of efficiency with height, as demonstrated by Sharp (1964) who placed four-foot high lucite rods in the Coachella Valley, California. After ten years he found that maximum abrasion occurred 25 cm above the ground. Maximum wind velocities increase above the ground, whereas maximum sand transport is near the ground. Maximum abrasion occurs where an intermediate number of particles are transported at intermediate velocities (Fig. 3.21).

The variation of sediment yield in the USA under natural vegetation and the variation of drainage density with climate show that maximum efficiency of fluvial processes is not in areas of maximum precipitation, but rather, it occurs in semiarid regions (Fig. 3.17). In arid regions low precipitation produces little runoff, and therefore, sediment yields are low in spite of abundant sediment available for transport. It is a transport limited situation. In humid regions the dense vegetation reduces the supply and size of sediment to the stream, and a supply-limited situation pertains. In semiarid regions runoff is sufficient to transport the supply of sediment from sparsely vegetated hillslopes and valley floors.

It is logical to assume that as gravitational forces increase on steeper slopes and as protective vegetation cover decreases that erosion rates will increase. However, in the first case as a slope steepens it receives less precipitation per unit area and this actually reduces raindrop impact and runoff erosion on slopes steeper than 45 degrees.

This can be expressed by the product of the sine (gravitational force) and the cosine (precipitation effect) of the angle of inclination (Fig. 3.22(*a*)). This theoretical relation is supported by rainfall–erosion experiments on slopes of less than 40 per cent (Fig. 3.22(*b*)); however, on steep slopes mass movement processes will probably dominate.

It is also logical to assume that erosion rates increase exponentially as vegetation cover decreases on a slope. However, experimental studies on a 10 per cent slope show that with less than about 15 per cent vegetational protection, in this case grass, sediment yields from the slope becomes essentially constant (Fig. 3.23). It

Figure 3.19. Illustrations of how efficiency of sediment transport (curve c) depends upon rate of sediment movement with increasing discharge (curve a), frequency of flows (curve b), and sediment size (t_1, t_2). (*a*) Sediment movement commences at low discharges and at frequent flows (t_1). Maximum work done (sediment transport) and maximum efficiency are at moderately frequent discharges (arrow). (*b*) Sediment movement commences at high discharge and infrequent flows (t_2); therefore maximum work done (sediment transport) is during large infrequent flows (arrow) (after Wolman and Miller 1960, from Baker, 1977).

appears that less than about 15 per cent vegetation cover is not effective in retarding raindrop impact and runoff erosion (Rogers, 1989).

In conclusion, it again becomes clear that an understanding of the processes operating and the material being affected by the process is necessary for a confident extrapolation.

Multiplicity

Definition

Multiplicity relates to diversity and to the condition of being various or manifold. In this discussion it refers to multiple causes acting simultaneously and in combination to produce a phenomenon. It differs from divergence as a single variable equation differs from a multiple correlation, and it can be represented by the composite hypothesis diagram of Fig. 2.2.

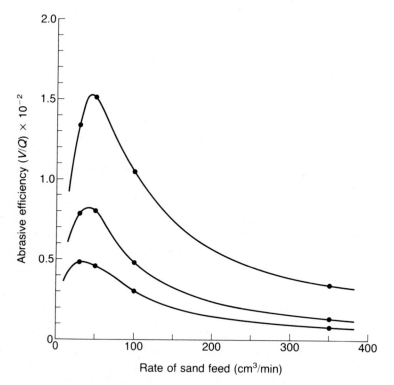

Figure 3.20. Relation between the abrasive efficiency of sand projectiles (V/Q), and the rate of sand feed, Q, for three different rocks (from Suzuki and Takahashi, 1972).

70

Problems of cause and process

Statement of problem

When dealing with complex systems, single explanations in most cases will not be sufficient, and therefore, a multiple explanation approach should be applied to problems. This is what was advocated earlier during the discussion of multiple working hypotheses (Figs. 2.2, 2.7).

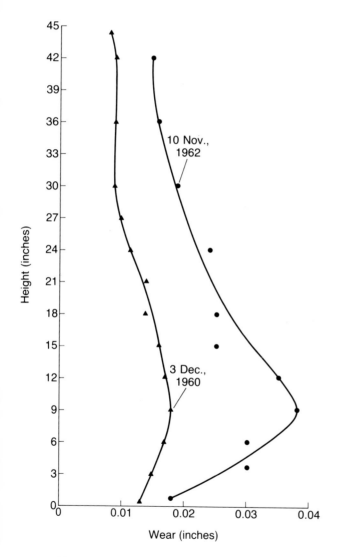

Figure 3.21. Wear by sand blast on a lucite rod (from Sharp, 1964).

Examples

Mosley and Zimpfer (1976) argue that a 'study using one approach provides only a partial explanation of the particular phenomenon under consideration'. Each variable provides only partial explanations of a given phenomenon, and therefore, one variable cannot provide a full understanding of most natural phenomena. They use studies of river meandering as examples, and complain that Leopold and Wolman's (1957) statistical study of the river meanders concentrates on river discharge and ignores other factors. Shen and Komura (1968) suggest how flow patterns may modify meander geometry, but they ignore differences in bank sedi-

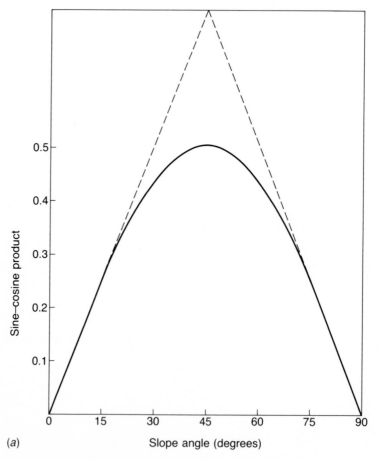

(a)

Figure 3.22(*a*). The sine–cosine product for slopes between horizontal and vertical.

ments, which can exert a major influence. On the other hand, Fisk's (1947) study of Mississippi River history stresses the importance of historical events in controlling river pattern, floodplain composition and river morphology, but it ignores many other factors such as sediment loads and hydraulics. Each study adds to the understanding of meandering, but they provide only a partial explanation of the phenomenon. In order to achieve a complete explanation of meandering or of any other natural phenomenon, it is necessary to combine partial explanations in a variety of forms. This procedure differs from that of multiple parallel hypotheses, in which one of several hypotheses is selected, and the remainder are discarded as

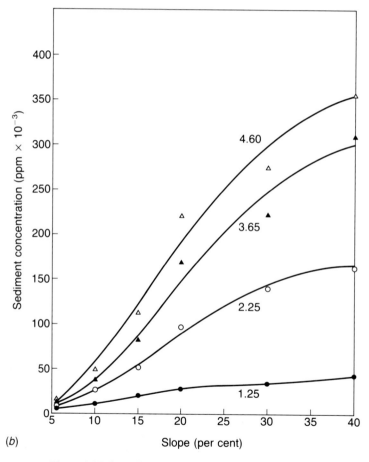

(*b*)

Figure 3.22 (*b*). Relation between sediment concentration in overland flow from slopes between 5 and 40 per cent. Numbers above curves indicate precipitation intensity in inches/hour (from Kilinc and Richardson, 1973).

73

invalid (Fig. 2.2(*c*)). In Mosley and Zimpfer's approach, there is no 'best explanation, because each explanation or variable used deals with a different aspect of the problem, such as historical evolution, relationship to contemporary physical processes, or significance of climatic and geologic factors'. Each explanation, therefore, provides only a partial explanation of a given phenomenon, which suggests an analogy with the statistical technique, analysis of variance (Mosley and Zimpfer, 1976, p. 386), and composite explanations are needed (Fig. 2.2(*d*)).

A discussion of river patterns can take the form of a purely modern problem with explanations based upon hydraulics and sediment load and transport, or it can take the form of a historical discussion of the cause of valley dimensions and slope, which clearly play a role in present patterns (see the section on time, above). The causes are partly historical and partly modern, and in each component there are multiple variables (Fig. 3.10).

In any historical explanation of meanders there must be consideration of paleo-discharges, sediment loads and neotectonics, which determine valley floor gradients (Fig. 3.18). The river flows on this historically determined valley slope, but modern stream gradient is determined by water discharge, sediment load, sediment type and hydraulics. For this example it is necessary to deal with both remote causes (historical) and proximate causes (modern) (Coyne, 1984, p. 129).

The multiple causes have different degrees of relevance to an effect. For example, I am allergic to dog dander and housedust (a cotton-decomposition product). I own a dog and in spite of my wife's best efforts there is dust in the house. These are the primary causes of my allergy problem, but on the majority of days I am not bothered by it. However, if humidity is very low or trees or weeds are pollinating I do have a problem. The low humidity and pollen are secondary causes of my allergy because, except perhaps under extreme conditions, without the dog and dust there would not be a problem. It would be easy to attribute all of my allergy problems to dog and dust, but I know that other secondary factors influence the situation. Hence, some secondary variables may be significant at certain times and at some locations but not always and not at other locations.

A final example is provided by the debate about Cretaceous extinctions. Hallam (1987, p. 1240) argues that the impact hypothesis does not explain the 'gradual or stepwise extinction patterns of many groups of organisms'. He notes that many investigators 'prefer some kind of extinction scenario completely excluding any extraterrestrial influence, one based on the combined effects of sea-level and climate change with volcanism on a massive scale". Here is multiplicity on a grand scale, and it is a repeat of many geologic controversies with the resolution of a controversy being a compromise which involves both explanations. Perhaps the Cretaceous extinctions were underway, as a result of long-term climate changes, sea-level lowering, volcanic effects, and then were exacerbated by an extraterrestrial event (Keller, 1989).

In conclusion, if there is more than one cause of a phenomenon, unless all are comprehended, extrapolation will be weak, and composite explanations are needed.

PROBLEMS OF SYSTEMS RESPONSE

Singularity

Definition

Singularity is the condition, trait or characteristic that makes one thing different from others. However, singular is not a synonym for unique. Humans are much the same, but each person has different characteristics that are singular. Landforms (rivers, hillslopes), when examined in detail, have sufficient differences so that they can be considered to be singular. Singularity is really the randomness or unexplained variation in a data set (Mann, 1970), and it is called indeterminacy by Leopold and Langbein (1963, p. 190).

Statement of problem

All earth scientists, after explaining a situation at location X, have been told by a colleague, 'Well, it isn't that way at location Y'. Of course, it is different, but that does not mean that generalizations obtained from studies at X are not applicable

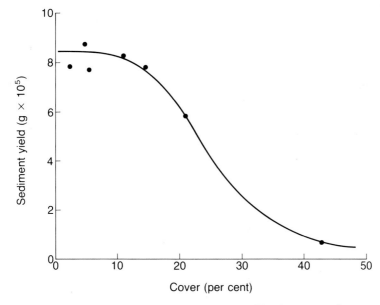

Figure 3.23. Relation between sediment yield and vegetational cover on a 10 per cent slope after the application of precipitation for 210 minutes (from Rogers, 1989).

75

at Y. They just may be obscured by singularity. That is, we can predict for a population based upon a large sample, but only order of magnitude estimates can be made for an individual.

If landforms are singular then each should respond to a change in either slightly or significantly different ways and at different rates. This really is the key to the difficulty of short-term prediction. General relationships (laws) may be of little value in predicting short-term changes. For example, when a geomorphic variable is plotted against a controlling variable, the data will usually scatter over half a log cycle or even over a full log cycle. This is a poor basis from which to predict individual response, and individual landform response to change may appear to be random. There can be singularity both in location and through time.

Examples

All field data and even experimental data from tightly controlled experiments show scatter. Each data point either falls close to a regression line or some distance from it. If meander characteristics (wavelength, amplitude) are measured along a river or if first-order channel length and drainage area are measured in a region of homogeneous lithology, climate, and relief the frequency distributions will probably be log–normal, with a large standard deviation. Thus, each meander and first-order basin will be singular in location.

Another example of singularity in location is provided by Fig. 3.24, which shows a plot that relates rock creep on a shale hillslope to slope inclination. The plot shows the expected positive relation between the two variables but the scatter is large, although mean values for 10 degree segments of the data fall very close to the regression line. Probably much of the scatter could be removed by determining soil characteristics at each measurement site, by measuring the exact slope upon which the rock rests rather than a 0.3 m slope segment, by determining the nearness of each rock to a plant and the possible shading of the rock by plants, and by determining the effect of animals (rodents, sheep, geomorphologists) on rock movement. Undoubtedly this additional information would decrease singularity and improve the plot, but is it worth it? Obviously, the answer depends upon the objective of the study. In most cases probably it is not worth it in time and cost. In the earth sciences it is often too time consuming and expensive to refine an investigation to the point that all randomness or singularity is explained, and therefore, the investigator must decide when to stop his research. However, when more data became available, situations that appear to be largely random (high degree of singularity) may, in fact, be largely deterministic. For example, if asked to predict how a drainage network will develop on a sloping surface one might try to determine what relation exists between channel length and drainage area (drainage density) for that surface and then fill the available space with the appropriate length of channels in a dendritic pattern. However, a detailed survey of the surface would undoubtedly reveal irregularities that will channel the flow, and

76

therefore, control the emerging pattern. Figure 3.25 shows exactly this situation for a small experimental drainage pattern. The main channels did follow irregularities that concentrated water, and, therefore, prediction of major pattern elements would have been possible in spite of what appeared to be wholly random initial conditions.

An example of singularity in time is the downstream shift of a meander. If a

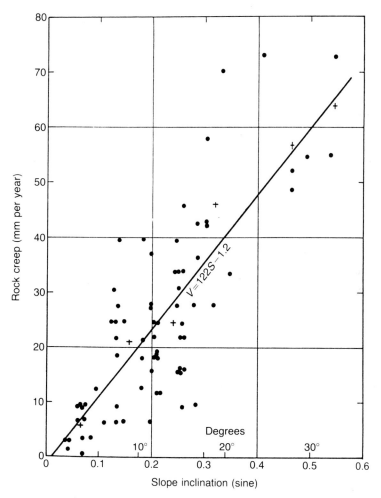

Figure 3.24. Relation between the sine of slope inclination and the rate of movement of rock fragments on Mancos shale hillslopes. Average rates of rock movement (V) for 0.10 increments of the sine of slope inclination (S) are shown as crosses (from Schumm, 1967).

bridge is present downstream the highway engineer will be concerned about the rate of movement of the meander toward the structure and the need to provide protection against it. A study of the morphology of the meander and a study of old maps and aerial photographs may indicate that, during the past one hundred years, the meander has shifted downstream at a rate of one meter per year. If the meander is a kilometer above the bridge, presumably the site will be safe for one thousand years. If the rate of movement appears to be on the order of two hundred meters per year there will be problems in five years. However, during the next year, that meander may encounter very resistant alluvium in the flood plain or buried bedrock that causes it to cease its downstream movement. On the other hand, a series of major floods may cause a great acceleration of meander shift, or it may cut off, removing the problem. thus, there is need to specify that the prediction concerning meander movement is a 'normic statement' that it is based upon normal circumstances, and any variability in the controls will cause a change in rate and perhaps even in the morphologic characteristics of the feature being investigated.

In conclusion, uncertainty of predictions pertains to all sciences, but accurate prediction in physics and chemistry are based upon large 'clean' samples, whereas earth-science samples are usually small, and because of the nonhomogeneity of each component of the sample, 'each may be considered singular if not unique' (Nairn, 1965). There is singularity of form and process in location and time, which can make extrapolation for singular features of the landscape or singular events very difficult.

Sensitivity

Definition

Sensitivity refers to the propensity of a system to respond to a minor external change. The change occurs at a threshold, which when exceeded produces a significant adjustment. If the system is sensitive and near a threshold it will respond to an external influence, but if it is not sensitive it may not respond. Sensitivity could be considered to be an aspect of singularity, but its significance requires separate treatment (Brunsden and Thornes, 1979; Wright, 1984).

Statement of problem

Owing to the complexity of Quaternary climatic and tectonic histories, topographic and stratigraphic variability can be conveniently explained as a result of climatic and tectonic events. In this way the compulsion to fit geomorphic and stratigraphic details into Quaternary chronology can be satisfied, as well as the basic scientific need to identify cause and effect. However, as the details of Holocene stratigraphic and terrace chronologies are studied, a bewildering array of changes are required to explain the behavior of rivers and drainage systems. In fact, it is accepted that some major erosional adjustments can be induced by rather insignificant changes in the magnitude and frequency of storm events (Leopold, 1951).

Another aspect of the problem is that within a given region all landforms may not have responded to the last external influence in the same way, and some may not have responded at all. This is a major geomorphic puzzle that is commonly ignored. If land systems are in dynamic equilibrium, components of the system should respond in a similar way to an external influence. Hence, the effects of hydrologic events of large magnitude should not be as variable as they appear to be. Mass movement, gully formation, and changes of river pattern may be triggered by relatively minor changes of a controlling variable if the geomorphic conditions have developed to conditions of incipient instability, as a threshold is approached (Schumm, 1977; Brunsden and Thornes, 1979; Wolman and Gerson, 1978). Therefore, a minor input may cause a major change, which elsewhere may have little or no effect (Begin and Schumm, 1984). A local change, if it left a stratigraphic record, could be interpreted to be the result of a major climatic or baselevel change, which would be incorrect.

The positive aspect of the problem is that if sensitive landforms (threshold

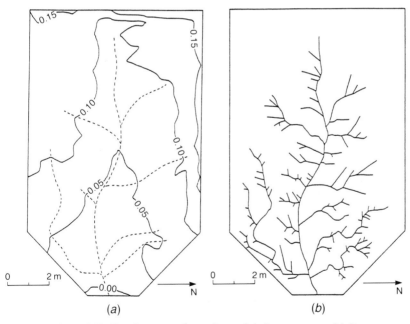

Figure 3.25. Development of experimental drainage patterns. (*a*) Contour map of initial surface. Dashed lines show initial drainage network. Contour interval is 0.05 m. (*b*) Fully developed drainage network after 1022 mm total precipitation (from Schumm *et al.*, 1987).

conditions) can be identified, they can be treated, and preventative conservation like preventive medicine can be a reality.

Examples

A good chemical example of sensitivity occurs where a solution slowly approaches saturation until a compound is abruptly precipitated. If the increase of concentration is the result of the addition of more of the compound to the solution then the control can be considered to be external to the system, and an extrinsic or allogenic threshold is exceeded. However, the slow evaporation or cooling of the liquid will bring the solution to a point of precipitation without an external influence, and an intrinsic or autogenic threshold is exceeded. Thresholds have been recognized in many fields and their importance in geography has been discussed in detail by Brunet (1968). Perhaps the best known to earth scientists are the threshold velocities that are required to set in motion sediment particles of a given size. With a continuous increase in velocity, threshold velocities are encountered at which movement begins, and with a progressive decrease in velocity, threshold velocities are encountered at which movement ceases.

Hydraulic thresholds are described by the Froude and the Reynolds numbers, which define the conditions at which flow becomes supercritical and turbulent. Particularly dramatic are the changes in bed-form characteristics at threshold values of stream power.

In most of the examples cited above, an external variable changes progressively, thereby triggering abrupt changes or failure within the affected system. Response of a system to an external influence occurs at an extrinsic threshold. That is, the threshold exists within the system, but it will not be crossed and the change will not occur without the influence of an external variable. However, as noted above, there are intrinsic thresholds, and changes can occur without a change in an external variable. An example is long-term progressive weathering under uniform conditions that reduces the strength of slope materials until eventually there is slope adjustment (Carson, 1971) and mass movement (Kirkby, 1973).

Probably the most common geomorphic example of sensitivity and an intrinsic threshold is the cutoff of a river meander. The meander develops to an unstable configuration through time, and it cuts off, only to form again.

Another example of an intrinsic geomorphic threshold resulted from field work in western Colorado (Patton and Schumm, 1975). In this area discontinuous gullies occupied the valley floors of some valleys. These gullies developed on steeper reaches of the valley floor. These steeper reaches form when alluvium accumulates and forms a convexity on the valley profile (Schumm and Hadley, 1957). In order to determine if threshold conditions of valley-floor gradient could be identified, the drainage area above each gully was measured on maps, and critical valley slopes at the point of gully development were surveyed in the field. No hydrologic records exist, so drainage-basin area was selected as a substitute for runoff and flood discharge. When valley slope is plotted against drainage area, the relationship is inverse (Fig. 3.26), with gentler slopes being characteristic of large drainage areas. As a basis for comparison, similar measurements were made in ungullied valleys, and these data are also plotted on Fig. 3.26. For the most part, the critical slopes of the unstable valleys coincides with the higher range of slopes of the stable valleys. In other words, for a given drainage area it is possible to define a critical valley slope above which the valley floor is unstable.

However, note that the above relationship does not pertain to drainage basins smaller than about four square miles. In these small basins, variations in vegetative cover, which are perhaps related to the aspect of the drainage basin or to variations in the properties of the alluvium, prevent recognition of a critical threshold slope. For areas larger than four square miles there are only two cases of ungullied valley floors that plot above the threshold line, and one may conclude that these valleys are incipiently unstable and that eventually a flood will cause erosion and trenching in these valleys.

Using Fig. 3.26, one may define the threshold slope above which trenching or

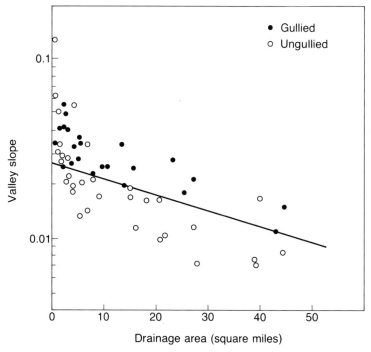

Figure 3.26. Relation between valley slope and drainage area, Piceance Creek Basin, Colorado. The line defines the threshold slope that separates gullied from ungullied valley floor. (From Patton and Schumm, 1975.)

81

valley instability will take place in the Piceance Creek area. This has obvious implications for land management for, if the slope at which valleys are incipiently unstable can be determined, corrective measures can be taken to artificially stabilize such critical reaches, as they are identified.

Note also that both intrinsic and extrinsic threshold conditions can be illustrated on Fig. 3.26. The local steepening of the valley slopes by deposition, which forms a valley fan or convexity, will result in the exceeding of an intrinsic geomorphic threshold, whereas an increase of runoff would have the effect of lowering the threshold line, which would put some valley floors that are now ungullied at risk. Begin (Begin and Schumm, 1979) reanalysed these data and concluded that it is more likely that a threshold zone exists rather than a sharp threshold line, and considering the effects of climatic fluctuations and the variability of erodibility of even similar materials this is a reasonable conclusion.

If on an alluvial fan the flow is spread across the apex, deposition will steepen

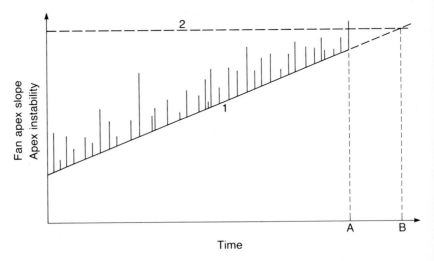

Figure 3.27. Relation through time between gradient at a fanhead and fan apex instability. Line 1 portrays the gradually increasing slope of the fanhead. When the ascending line of fanhead slope intersects line 2, which represents the maximum slope at which the apex is stable, trenching will occur, at time B. Superimposed on line 1 are vertical lines representing changes in fanhead instability that are related to high-magnitude runoff events or longer-term climatic fluctuations. Normally, the operation of these processes has little significant morphologic effect on the alluvial fan. However, when the fan slope and apex instability are high, trenching will occur sooner than expected (at time A) when a large-magnitude event exceeds the stability threshold (line 2). In reality, the event merely precipitated the eventual incision at time A rather than at time B (after Schumm and Hadley, 1957).

the fanhead until a geomorphic threshold is exceeded and a fanhead trench forms (Fig. 3.27). During experimental studies of this phenomenon, the gradient of the fan apex increased by deposition, and fan apex instability increased. The critical slope for entrenchment on the experimental fluvial fan was about 2.7 degrees (Schumm *et al.*, 1987); it would, of course, be different under other climatic, hydrologic, and sedimentologic conditions. Changes in these factors might change the critical slope for entrenchment or modify the landform so that its stability threshold is changed. Thus, for example, tectonics may increase slope beyond its threshold value and initiate trenching. Tilting might also reduce slope, increasing stability and promoting apex aggradation. On the other hand, climatic change may raise or lower the threshold of stability by changing sediment yields, flood magnitudes, vegetational characteristics, and overall erodibility, so that a new threshold slope is established that may be either closer to or farther from the actual value of slope on a fan apex (Fig. 3.28).

Recognition of explanations of the importance of the slope threshold permits development of explanations of alluvial fan behavior that do not rely on changes

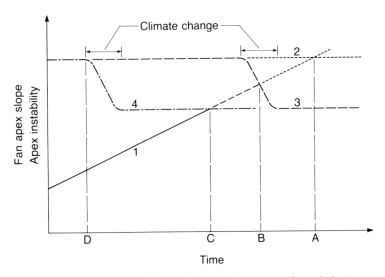

Figure. 3.28. Effect of climate change on the timing of morphologic response. Line 1 depicts the increase of fan apex slopes with time. Under unchanging conditions (line 2) fanhead slopes exceed their stability threshold at time A and incision commences. Climate change, which in this case establishes a new, lower threshold of stability (line 3), triggers incision at time B. If the same climate change was initiated much earlier, at time D, when the fan displayed greater inherent stability, then fanhead trenching at time C appears essentially unrelated to the climate change depicted by line 4 (from Schumm *et al.*, 1987).

83

of external conditions. This is not to deny the significance of such factors as tectonism or climatic change, but it also permits an explanation of differences in the timing, magnitude, and direction of responses of alluvial fans to the same external change, in terms of their relative stability (Fig. 3.29). The experimental results indicate that fan behavior may be directly explicable in terms of a threshold model.

Within a landscape composed of singular landforms there will be sensitive and insensitive landforms. The sensitive landforms will respond significantly to even a minor change but certainly to a large hydrologic event (Begin and Schumm, 1984). For example, sediment may accumulate in a valley or channel until it had reached a gradient threshold of instability and it is trenched or flushed during an apparently normal hydrologic event. In a series of meanders one may grow in amplitude to a condition where cutoff is inevitable. That meander was sensitive. The others were not, although all were singular.

Experimental studies of river pattern show that as flume slope and sediment loads increase, the pattern changes from straight to meandering to braided. The inception of sinuosity and braiding takes place at a limited range of slopes. For example, at locations b, c and d on Fig. 3.18 there are river pattern or sinuosity thresholds, obviously a channel that plots on this diagram near b, c or d is sensitive. Those plotting near a or e are insensitive to pattern changes.

The climatic changes of the Pleistocene–Holocene transition have greatly affected river patterns, as studies in many locations have demonstrated. However, as within any region there are sensitive and insensitive river patterns and, therefore, although the extrinsic change was large, the response of some rivers was long delayed.

The Polish Plain provides an example of this type of change (Mycielska-Dowgiallo, 1977; Froehlich *et al.*, 1977; Starkel, 1983) as Pleistocene braided rivers changed to meandering (Fig. 3.30). A transition from large meandering channels to the present river conditions can be documented by a decrease in size of the meanders that were preserved by cutoffs, although in some areas the modern rivers show a tendency to braid again as a result of agricultural activities, deforestation, and other factors.

The results of studies of six Polish rivers show that the change from braided to meandering took place between 13 000 and 9000 years ago and in some cases perhaps as late as 6000 years ago (Fig. 3.31). This lag indicates a great range of channel sensitivity, and it indicates how variable river response to climate change can be a further example of divergence.

In conclusion, the recognition of sensitive threshold conditions appears to be essential in order that reasonable explanations and extrapolations can be made in geomorphology, soil science (Muhs, 1984), sedimentology and stratigraphy (Anderson, 1986), and many environmental and ecosystem areas (Westman, 1978; May, 1977).

Complexity

Definition

When something is complex, it is composed of numerous interconnected parts. Natural systems are inherently complex, but the complexity referred to here is the complex response that results when the system is perturbed. The complex system when interfered with or modified is unable to adjust in a progressive and systematic fashion, and its response can be complex.

Statement of problem

Obviously with complex systems extrapolation is very difficult. As an example, Weinburg and Weinburg (1979, pp. 221, 312) conclude that if one cannot think of three things that can go wrong then one does not understand the system. Thomas (1979, p. 110) puts it differently by pointing out 'You cannot meddle with one part of a complex system from the outside without the almost certain risk of setting off disastrous events that you hadn't counted on in other, remote parts.

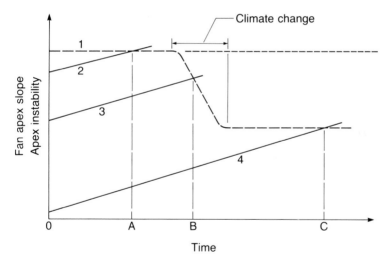

Figure 3.29. Effect of climate change on a group of alluvial fans. At time 0, the alluvial fans display contrasting degrees of inherent stability under a specific climatic and tectonic setting. One fan, represented by line 2, exceeded its stability threshold without external change and fanhead trenching occurred at time A. Climate change prematurely triggered incision at time B on another fan (line 3), which would have been stable at time B had conditions remained unchanged. Finally, the fan displaying the greatest relative stability (line 4) at time 0 showed considerable lag by responding to the climate change at time C, well after the establishment of new conditions (from Schumm *et al.*, 1987).

If you want to fix something you are first obliged to understand, in detail, the whole system.' This has been amply displayed by the effects of channelization of streams (Schumm *et al.*, 1984), dam construction on rivers (Williams and Wolman, 1984) and beach stability (Griggs, 1987). Of course, success becomes even more difficult when it is necessary to do more than describe. Therefore, prediction is difficult and prediction can be totally misleading if instead of a progressive adjustment to change a complex sequence of events is the result.

Figure 3.30. Paleochannels in the Prosna valley at Mirków, Poland. 1. high Pleistocene terrace with braided-channel pattern; 2. and 3. slope to valley floor; 4. modern Prosna River; 5. braided pattern on valley floor; 6. paleomeanders; 7. point bars (from Kozarski and Rotnicki, 1977).

Examples

A complex system, when interfered with or modified, may not adjust in a progressive and systematic fashion (Schumm, 1977). As an example, consider rejuvenation of a drainage system. As the effects progress upstream, downstream reaches are affected long before the upstream reaches (problem of location). This creates a situation in which it is very difficult for a given reach of a channel or a given tributary to adjust progressively. In fact, there will be hunting for a new condition of stability, which is referred to as complex response (Fig. 3.32(a)). In high-energy systems the behavior can even be episodic with periods of aggradation interrupting degradation (Fig. 3.32(b)) until a new condition of stability has been achieved. This produces a very complex geomorphic and stratigraphic record, the details of which cannot be attributed to external influences but rather to the adjustment of the system itself.

For example, incision of streams crossing the Canterbury Plain in New Zealand into glacial outwash and incision of Sierra Nevada streams into hydraulic mining debris has produced 'degradational terraces'. These unpaired terraces reflect pauses in downcutting, as the channels became clogged with sediment. Some tributaries of the Bear and Yuba Rivers in California have as many as 11 unpaired terraces (Fig. 3.33), all of which formed since 1880 (Wildman, 1981; see Schumm

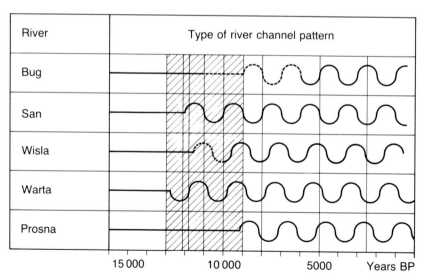

Figure 3.31. Change of river channel patterns on the Polish Plain at the Pleistocene–Holocene transition. The straight line represents braiding, curved lines represent meandering, the dotted line represents the assumed river condition. The cross-hatched zone represents the period of river change on the Polish Plain (from Kozarski and Rotnicki, 1977).

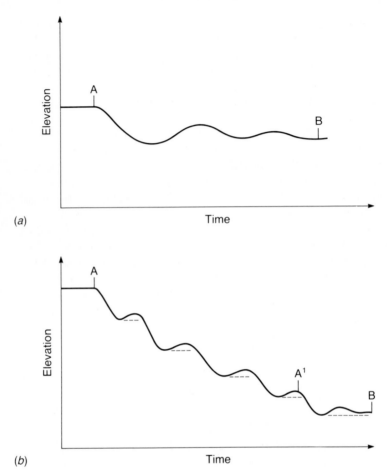

Figure 3.32. Diagrams illustrating (*a*) complex response and (*b*) episodic erosion. In each case a stream is affected at time 'A' by a climate, baselevel or land–use change that induces degradation. When the impact is relatively small (*a*), the stream degrades, aggrades, and degrades until a new condition of relative stability is achieved at time 'B'. When the change is large (*b*), and large quantities of sediment are moved, degradation is episodic being interrupted by periods of aggradation until relative stability is achieved at time 'B'. Area above dashed lines indicates extent of aggradation. The complex response of (*a*) occupies the space between 'A[1]' and 'B' on the diagram of episodic erosion (*b*) (from Schumm, 1977).

et al., 1987, p. 125). These are only landscape details, but careful evaluation of a responding stream is required because aggradation may follow the expected degradation, as a natural result of increased sediment movement from upstream (Womack and Schumm, 1977; Boison and Patton, 1985; Waters, 1985).

During experimentation, a small (10 by 15 m) drainage system was rejuvenated by a 10 m change of baselevel. As anticipated, baselevel lowering caused incision of the main channel and development of a terrace (Fig. 3.34(*a*)). Incision occurred first at the mouth of the system, and then progressively upstream, successively rejuvenating tributaries and scouring alluvium previously deposited in the valley (Fig. 3.34(*b*)). As erosion progressed upstream, the main channel became a con-veyer of upstream sediment in increasing quantities, and the inevitable result was that aggradation occurred in the newly cut channel (Fig. 3.34(*c*)). However, as the tributaries eventually became adjusted to the new baselevel, sediment loads decreased, and a new phase of channel erosion occurred (Fig. 3.34(*d*)). Thus, initial channel incision and terrace formation was followed by deposition of an alluvial fill, channel braiding, and lateral erosion, and then, as the drainage system achieved stability, renewed incision formed a low alluvial terrace. This low surface formed as a result of the decreased sediment loads when the braided channel was converted into a better defined channel of lower width–depth ratio.

The experimental results indicate that an event causing an erosional response within a drainage basin (tilting, changes of baselevel, climatic and/or land use)

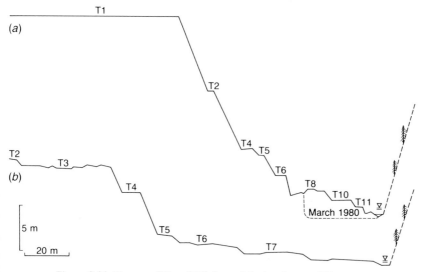

Figure 3.33. Terraces (T1 to T11) formed during the past 100 years as sediments produced by hydraulic mining were removed from the Bear River valley, Sierra Nevada, California, (*a*) Gas Canyon incision, 21 m deep, (*b*) above Red Dog Ford incision, 12.5 m deep (from Wildman, 1981).

automatically creates a negative feedback (high sediment production) which results in deposition within the drainage network; this is eventually followed by incision of alluvial deposits as sediment loads decrease.

The complex response observed during experiments and in the field is apparently a quest for a new equilibrium by a complex system. However, the changes documented in Fig. 3.34 are minor, compared to those accompanying major periods of aggradation or degradation, which result from major climatic or baselevel changes.

It is well established in the geomorphic literature that a major reduction of baselevel will cause progressive downcutting and readjustment of the stream gradients until a new graded or equilibrium situation has been developed. In fact, the fluvial system may not be capable of degrading in this way when sediment movement is great (Hey, 1979).

Where major incision has occurred in alluvium and relatively weak rocks, evidence of pauses in the erosional downcutting are found, but this is usually attributed to some external influence, such as variations in climate, the rate of baselevel change, or variations in the rates of uplift of the sediment source area. How-

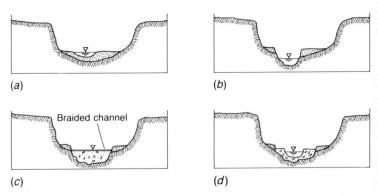

(a)   (b)

(c)   (d)

Figure 3.34. Diagrammatic cross sections of experimental channel 1.5 m from outlet of drainage system showing response of channel to one lowering of baselevel (from Schumm and Parker, 1973). (*a*) Valley and alluvium, which was deposited during previous run, before baselevel lowering. The low width–depth channel flows on alluvium. (*b*) After baselevel lowering of 10 cm, channel incises into alluvium and bedrock floor of valley to form a terrace. (*c*) Following incision, bank erosion widens channel and partially destroys terrace. An inset alluvial fill is deposited, as the sediment discharge from upstream increases. The high width–depth ratio channel is braided and unstable. (*d*) A second terrace is formed as the channel incises slightly and assumes a low width–depth ratio in response to reduced sediment load. With time, in nature, channel migration will destroy part of the lower terrace, and a flood plain will form at a lower level.

ever, studies in the Douglas Creek drainage basin of western Colorado support the idea of discontinuous downcutting. The investigation of recent erosional history of this valley shows that modern incision of the valley fill began after 1882. Yet, there are four surfaces now present below the two pre-1882 surfaces (Fig. 3.35). These surfaces are unpaired, discontinuous terraces that elsewhere have usually been explained by the shifting of a channel laterally across the valley floor, during progressive downcutting (Davis, 1902). In the Douglas Creek Valley, however, downcutting was discontinuous. In fact, during pauses in downcutting there was deposition. The denudation scheme as sketched on Fig. 3.35 portrays the sequence of events in this 100 km² drainage basin in western Colorado. That is, during incision of the main channel, there is rejuvenation of tributaries and a progressive increase in sediment yield from upstream. Sediment loads become so great that downcutting ceases and deposition begins. Deposition continues until it is possible for the channel to incise again and to continue the downcutting process.

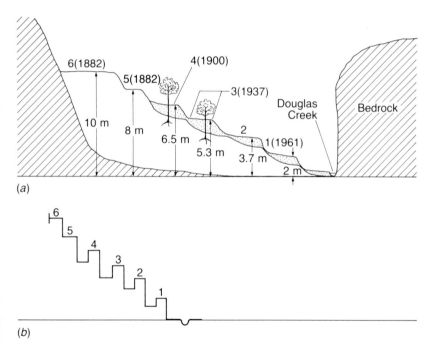

Figure 3.35. (*a*) Sketch of Douglas Creek valley showing erosion surfaces formed since 1882. Age of surfaces is based on tree-ring dating and historical data. Note burial of trees by deposition. Surfaces 5 and 6 were present before modern erosion began after 1882. (*b*) Summary of behavior of Douglas Creek. Vertical segments indicate incision or deposition, horizontal segments periods of relative stability (from Womack and Schumm, 1977).

91

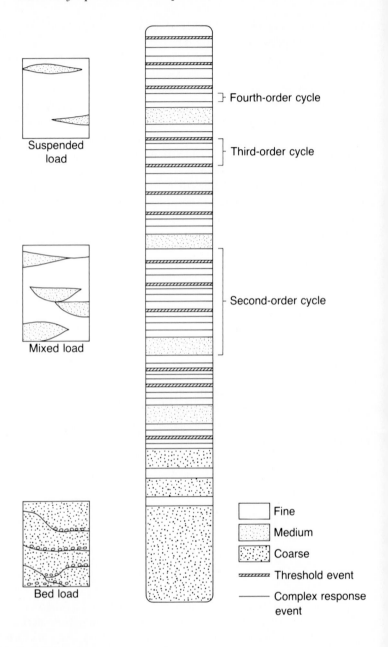

Other examples of this behavior have been described by Born and Ritter (1970) and by Gage (1970).

The temporary storage and removal of sediments should produce complexity in fluvial deposits (Dott, 1983), and cycles of several dimensions should be found within a fluvial depositional unit (Fig. 3.36). There could be at least five types of fining-upward cycles of decreasing magnitude in the deposits associated with erosional evolution of a landscape. The primary cycle is related to denudation following uplift, with maximum sediment production at the beginning of the erosional event and a progressive decrease in quantity and size of material through time to yield a massive fining-upward sequence. However, interruptions due to isostatic adjustment will produce higher sediment yields from the source area. Therefore, within the primary cycle associated with tectonics, there should be second-order cycles associated with isostatic adjustment and perhaps major climatic change. Between these events third-order cycles are related to the exceeding of geomorphic thresholds. These third-order cycles will be of much smaller dimension, and yet they should be important for the concentration of heavy minerals and the development of channelling in fluvial deposits. Fourth-order cycles will be related to the complex response of the fluvial system to any of the above changes, either tectonic, isostatic, climatic, or geomorphic threshold. These cycles of smaller dimension will result from the attempt of the system to adjust to changes related to the primary, secondary, and tertiary cycles. Finally, fifth-order cycles will appear that are related to the seasonality of hydrologic events or to major flooding, and in the stratigraphic record these will appear as thin fining-upward depositional units.

As denudation progresses the decreasing magnitude of sediment loads will cause changes in the types of rivers that bring the sediment to the depositional site (Fig. 3.36) from braided bed load to increasingly sinuous channels (mixed load and suspended load). At some point the threshold between braided and meandering will be crossed with a relatively abrupt change in the deposits associated with each river type (Fig. 3.18).

In conclusion, without an understanding of the response of complex systems to

Figure 3.36. Diagrammatic model of a major fining-upward sedimentary cycle that is related to uplift and the erosional evolution of the landscape. The major cycle is composed of second-, third- and fourth-order cycles that relate to isostatic adjustment (2nd-order cycle), geomorphic thresholds (3rd-order cycle) and episodic behavior and complex response events (4th-order cycle). During the primary cycle three types of channels will be functioning and the deposits associated with each are shown at appropriate places. In reality this simple and idealized model will be complicated by climate change and additional tectonic activity and even large hydrologic events (after Schumm, 1977).

change, extrapolations cannot be made with confidence, and indeed, the effects of human activities and the landscape cannot be predicted confidently.

DISCUSSION

A wide range of problems that are frequently encountered during research have been discussed in varying detail. I believe that at least seven of them, time, space, location, sensitivity, complexity, convergence and divergence, must be considered in any study of natural phenomena. The three problems, singularity, efficiency and multiplicity, probably warrant less concern because they are inherently accommodated by most investigators. Nevertheless, they deserve to be included in any discussion of problems of research. It has been claimed that convergence is really not a problem (Haines-Young and Petch, 1983) because detailed investigations will reveal differences in the landforms and the causes and processes that formed them. This is, of course, true and it is probably true of all of the ten problems. If studies are sufficiently detailed and if the data are sufficient then the problems may not exist, but they will exist prior to the detailed study. In fact, the point to be made is that a ruling hypothesis based upon reconnaissance or incomplete investigations must be avoided.

No attempt has been made to discuss solutions to each problem, as there are general solutions that can be applied to several problems, and these will be considered in the next chapter. Awareness that the problems exist is in itself a considerable aid to research success, and an investigator might consider maintaining a check list of the problems or keeping them in mind by the use of a mnemonic aid such as that presented in Table 3.1. Name magic may be more important than we think.

4 Scientific approach and solutions

I cannot give any scientist of any age better advice than this: the intensity of the conviction that a hypothesis is true has no bearing on whether it is true or not.
Medawar, 1979a, p. 39

Because of the diverse and multiple problems faced by earth scientists, we must abandon the concept of a single scientific method and instead consider a scientific approach. The approach is less involved with procedure and more with the state of mind of the investigator during research activities. It is unfortunate that there is no 'cookbook' for research, but if there existed a single method the pursuit of science would be considerably less exciting. Therefore, although a detailed step by step method is not available, the characteristics of a scientific approach can be listed. In the best of all possible worlds the scientific approach has five characteristics (Feigl, 1953).

1. The approach is systematic and orderly. There should be planning prior to the investigation in order to anticipate problems that will develop. One then attempts to proceed in an orderly fashion to collect relevant data. Obviously, although a great effort is made to be systematic and orderly, in most cases mistakes will be made. These are characteristic of research. Work in many cases will be repeated, as hypotheses are revised.
2. The approach is characterized by a lack of bias. The work must be carried out objectively. There will be no conflict of interest, and there should be no ruling hypothesis directing the research.
3. The approach will be characterized by absolute intellectual honesty. The results of the research will be reported accurately whether the results are negative or positive. There is no attempt to fabricate or manipulate the data. This means that the work can be evaluated and used by others, and the work of one scientist can be tested by another. In addition, the work of predecessors and colleagues must be credited. However, sometimes with the best of intentions the work of others will be ignored as Gilbert's (1904, pp. 115–16) 'case of plagiarism' illustrates (Fig. 4.1).
4. The approach is based upon physical and chemical laws and therefore, hypotheses can be and will be tested. Metaphysical explanations are not acceptable.
5. The goal of the approach is to develop broad generalizations in order to classify and to express relations, as quantitative models, with definiteness and precision. This involves identification of cause and effect if such is appropriate.

95

The search for scientific knowledge can be identified by at least the above criteria (Feigl, 1953). Therefore, the method of the search itself is less important than the characteristics of the approach. In fact, because of the great number of types of science it is only the scientific approach, with the above characteristics, that will be applicable to any science; whereas, a strictly defined scientific method will not. Bunge's list of the fifteen attributes of scientists (Table 2.3) conforms to the criteria of the approach listed above.

Because of the above characteristics, science is self-correcting (Hallam, 1983, p. V). Error or fraud will eventually be revealed with considerable satisfaction by a colleague or former student. Therefore, it is in the self-interest of the scientist to use the scientific approach. Self-correction ensures the advance of science, and it should enforce objectivity and honesty in research.

We can hope that the scientific approach will assist an investigator in avoiding harsh criticism of the type leveled by a committee of the Australian Parliament at researchers investigating the effects of the Crown of Thorns starfish on the Great Barrier Reef as follows (Ford, 1988):

> The Committee finds it difficult to understand why it appears some scientists refuse to consider rationally the views of other scientists or to modify their opinions in the light of new information . . . It also concerns the committee that some scientists have been so preoccupied with either advancing their own theories or rejecting the opinions of their opponents that some important developments appear to have been given insufficient attention.

Although the threat of such criticism is constraining it should not inhibit imagination. Many now accepted hypotheses were formerly considered to be 'outrageous' (e.g. continental shift, megafloods). Most were viewed as being catastrophic, and therefore, they were unacceptable. However, it is argued that these leaps of imagination can contribute importantly to the advance of science (Huggett, 1988), and therefore, they must be considered to be the basis for additional research, perhaps in new directions.

ESTABLISHING CAUSE AND EFFECT

Finding the cause of a phenomenon is a major part of the scientific approach in earth science, as it is in many of the sciences. As we have seen, similar causes can have different effects (divergence) and different causes can have similar effects (convergence). Indeed, cause–effect relations may not be readily determined in some cases. Susser (1973) attempts to assist in the search for cause by providing a list of five criteria for establishing causal relations. He refers to these as 1. time sequence, 2. consistency of associations, 3. strength of association, 4. specificity of association, 5. coherent explanation.

Obviously, item 1, time sequence, requires that the cause precedes the effect, a common-sense criterion. Nevertheless, just because something precedes some-

---

DISCUSSION AND CORRESPONDENCE.

A CASE OF PLAGIARISM.

To THE EDITOR OF SCIENCE: In a note on 'The Mechanism of the Mont Pelée Spine' (SCIENCE, June 17, 1904), I say: 'So far as the literature has come to my attention, it has failed to include a factor which appears to me of prime importance,' etc. Through this sentence I claim originality, and presumptive novelty, for an idea which I now know not to have been novel, and think not to have been original. The idea was published six months earlier by Dr. A. C. Lane in a note on 'Absorbed Gases and Vulcanism' (SCIENCE, December 11, 1903). It is not necessary, in dealing with my friend Dr. Lane, that I disclaim intentional plagiarism, but, as I find interest in the mental process of my blunder, I venture to relate what I suppose to be its history. It is altogether probable that I read Dr. Lane's note when it appeared, but the mental impression it made was so faint that in re-reading it now I can not definitely remember seeing it before. Nearly a half year later an idea as to the Pelée spine occurred to me and I wrote it out for publication. While I supposed the idea original, there was in my mind a faint suspicion that the suggestion might have come from some outside source, and this suspicion led me to search all the literature of the spine that I could recall having seen—but I did not recall that Dr. Lane had made a contribution. Thus a mental impression too faint for complete identification, now that attention is directed to it, nevertheless rose into consciousness with the semblance of a spontaneous idea, and gave rise to a distinctly plagiaristic publication.

G. K. GILBERT.

SAN FRANCISCO,
June 28, 1904.

Figure 4.1. G. K. Gilbert's case of plagiarism.

thing else is no reason to conclude that a causal relationship exists. Item 2 requires that there be a consistent relation between the causal variable and the effect and that in experimental work the relations hold during replication of the experiments. Item 3 requires that the relation between the variables be strong. That is, a high correlation coefficient results upon statistical analysis of the data. However, correlation, for example, the relation between skirt lengths and stock market values, does not always indicate causality. Item 4 requires that the use of the causal variable results in strong prediction. The ideal situation is a strong one to one relation between the two variables. That is, in a single case the prediction holds. Item 5 requires that a rational explanation of the relationship can be produced from known physical and/or chemical relations.

All the above can be termed a rational approach to a problem. Mackin (1963) made a strong case for development of understanding rather than simply reporting empirical relations that may or may not be meaningful and, indeed, can be misleading. For example, I once reviewed a manuscript that showed a strong negative relation between drainage density and relief. This, of course, makes little sense because higher relief means more potential energy and more channels should form instead of less. What was ignored was the effect of lithology and, of course, the multiple hypothesis approach (Fig. 2.2). Figure 4.2 shows how this empirical statistical approach can mislead.

Any of the ten problems (Chapter 3) that are considered to be relevant to a research problem must be resolved before a satisfactory explanation can be developed and certainly before extrapolation is attempted.

SOLUTIONS

Hypotheses are for testing, and indeed, without a hypothesis to test research will be empirical rather than rational (Mackin, 1948). The advantage of having a hypothesis is that it guides data collection. Hsü (1983) clearly demonstrates this, as he describes how drilling locations were selected during an oceanographic cruise in the Mediterranean basin. Depending on the hypothesis to be tested, the location of the next drilling site was selected with the hope that it would yield the required data. Of course, in each case the results can be positive or negative, and the hypothesis can be strengthened or weakened. In many cases, although the results are negative, the hypothesis cannot be falsified because the results are not conclusive. This, of course, will be a common result (Fig. 2.5).

A hypothesis must be testable. Each of the ten problems can be considered to be a hypothesis that needs to be tested and, indeed, for each investigation the relevant problems should be identified and given consideration. Rather than discuss individual solutions for each of the ten problems, it is perhaps more reasonable to group them and to suggest general solutions that can apply to several problems. The solutions are threefold as follows (Table 4.1):

1. assemble historical information, and develop a history of past events that can lead to prediction (Fig. 1.2);
2. develop an understanding of the processes operating, and determine the physical and chemical relations that are applicable;
3. compare in space, and determine the different characteristics that exist at different locations.

The first solution will answer the question 'What was it?' The second solution should answer the question 'What controls it?' The third solution will answer the questions 'How general is it?' and 'Where does it fit in the spectrum of this phenomenon?'

History

A historical background will assist in all problems, but it is most important for understanding the importance of time, location, sensitivity and complexity (Table

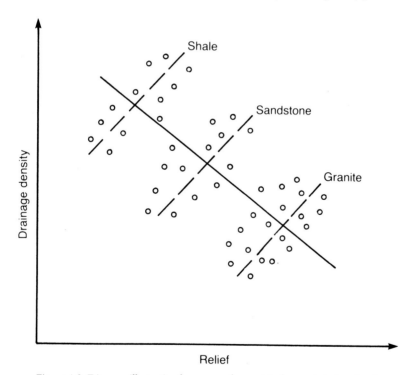

Figure 4.2. Diagram illustrating how a purely empirical or statistical evaluation of data can be misleading. It is rational to conclude that drainage density increases with relief for drainage basins developed on one lithology. Therefore, the statistically significant negative empirical relation (solid line) is replaced by positive relations (dashed lines) when the data are stratified by rock type.

Table 4.1. *Summary of solutions to problems of explanation and extrapolation*

History		Time
	⎰	Location
	⎱	Sensitivity
		Complexity

Process		Convergence
		Divergence
	⎰	Complexity
	⎱	Sensitivity
		Efficiency
		Multiplicity

Comparison		Space
		Location
	⎰	Singularity
	⎱	Sensitivity
		Complexity
		Multiplicity

4.1). Obviously the time problem requires historical information, and the problems of location, sensitivity and complexity can be solved by following a process or a change through time. In this way sensitive conditions can be recognized and a complex response documented. Also changes at one location through time can be used to predict changes at other locations, but a major difficulty may be obtaining the historical information.

The combination of a model of the present with historical information is of great value for prediction (Table 1.1). One example of the present–past type of approach was its use in order to aid highway engineers in the recognition of river hazards (Shen and Schumm, 1981). For the study of present conditions the collection of available topographic maps, aerial photographs, soil maps, land-use maps, as well as hydrologic and meteorological data, and information on the geotechnical properties of bank materials, bank vegetation and the hydraulic character of flow, is necessary. These types of information permit a description of the present situation, but such a glimpse at one instant of time does not provide an adequate basis for prediction of channel stability. However, a historical study of the river and its behavior may provide a basis for evaluating the relative stability of the channel. A key concept in historical research is that an understanding of past events provides a basis for prediction, but natural systems will behave in a predictable fashion only if there is no change in the variables that influence the system.

In any case, the documentation of past conditions is of considerable value in determining the potential for future problems. Information on the near past (Table 1.1), which can be used, for example, to determine the stability of a bridge crossing, can be obtained in at least five ways as follows:

1. Determine the history of nearby bridges. If the new bridge is to replace an older one, considerable information should be available on the past morphology and behavior of the river at that site. For example, channel width and the distance from the crown of the highway to the stream bed will be available. Any change can be readily determined by a comparison of the present cross-section characteristics with those at the time of the construction of the old bridge.

2. Conversations with long-time residents of the valley can be useful in establishing the relative stability of the river channel. Recollections are sometimes suspect, but old photographs of the river obtained from private collections, family albums and local historical societies can be invaluable. State archives and historical societies frequently contain photographs of old bridges and fords, and hence they are a source of valuable information.

3. In the mid-west and western United States, General Land Office surveys made in the nineteenth century frequently provide information on former river widths and patterns. The earliest maps can be compared with more recent topographic maps and aerial photographs. Aerial photographs may be available for several years after the early 1940s.

4. Records such as newspaper reports, railroad company files, church records, court transcripts, and accounts of early travelers are all possible sources for identifying possible channel changes.

5. Gaging station records can be used to assess channel stability and to detect long-term hydrologic trends or the importance of large floods.

Although the above procedures are aimed at very-short-term prediction, they can be applied to postdiction as well.

A longer historical record can be developed by using the location for time substitution (LTS). This has been used with great success to determine future changes of rapidly evolving landforms such as gullies, arroyos and channelized streams, and it can be used to determine long-term evolutionary changes of landscapes (Paine, 1985). Therefore, it could also be referred to as the location for evolution substitution.

For example, if a series of cross-sections are surveyed along a channel (Fig. 4.3) that has been incised, as a result of natural or human-induced changes (e.g. channelization) an evolutionary model of channel adjustment can be developed (Fig. 4.4). Glock (1931) used this technique to develop a model of drainage network evolution (Fig. 4.5). In spite of severe criticism of his model, subsequent experimental studies support his conclusions (Schumm *et al.*, 1987). The model presented in Fig. 4.4 was developed for incised channels in northern Mississippi, and it has both academic and practical value because it permits estimation of sediment production and agricultural land loss (Schumm *et al.*, 1984). The location for time substitution can be an effective means of developing a model of evolving landforms.

In using LTS it is important to compare features produced by the same processes that are operating under the same physical conditions. For example, the evolution of an incised channel in alluvium can be determined by surveying cross-sections at several locations where the channel is in alluvium (Fig. 4.3), but one cannot combine data or compare channels in weak alluvium with channels in resistant alluvium or bedrock and expect to find meaningful results. Nor can one develop a model of drainage network evolution (Fig. 4.5) if the lithology at each site is different.

Process

Obviously, the problems that are process controlled cannot be solved without an understanding of the processes operating. This involves not only a general under-

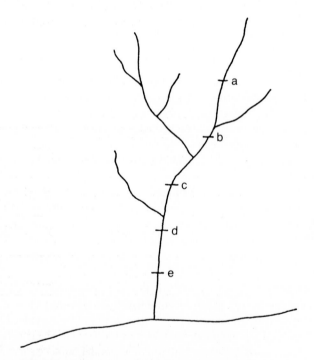

Figure 4.3. Sketch shows method used to obtain data for location for time substitution (LTS).

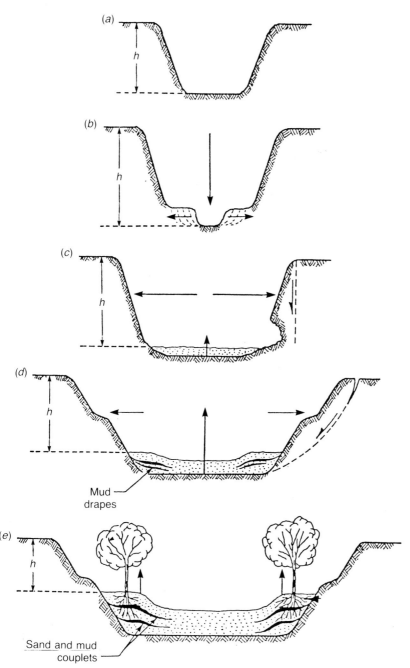

Figure 4.4. Evolution of incised channel from initial incision (a, b) and widening (c, d) to aggradation (d, e) and eventual stability (e) (from Watson and Harvey, 1988).

103

Table 4.2. *Mechanisms of slope segment production (from Schumm, 1966)*

I   Convex segments
   1. Erosion
      (a) creep
      (b) rainwash (on relatively permeable soils and upon moving from an upper flat segment to a lower steeper segment)
      (c) mechanical and chemical weathering of massive rocks
      (d) pressure-release jointing (formation of domes)
      (e) basal erosion (lateral stream erosion, spring sapping)
      (f) incision of widely spaced streams in areas of low or moderate relief
   2. Deposition
      (a) mantling of existing slopes with loess or volcanic ash
II   Concave segments
   1. Erosion
      (a) rilling (erosion greatest on middle segment)
      (b) creep (least efficient at slope base)
      (c) piping
   2. Deposition
      (a) accumulation of talus at slope base
      (b) accumulation of colluvium at slope base
      (c) accumulation of volcanic ejecta to form a cinder core
III Straight segments
   1. Erosion
      (a) rainwash
      (b) incision of closely spaced streams in areas of moderate or high relief
      (c) mass movement (separation along vertical joints including pressure-release jointing)
      (d) basal erosion (lateral stream erosion, wave action, erosion of weaker underlying rock)
   2. Deposition
      (a) talus, sand, volcanic ejecta at angle of repose

standing of processes (e.g. fluvial, aeolian, glacial, mass–movement), but also some knowledge of details of the process in order to recognize convergence and divergence and to determine what is involved in determining complexity, sensitivity, efficiency and multiplicity (Table 4.1). For example, convex hillslopes are formed by creep, the slow downslope movement of soil by gravity, but they can also be formed by rapid stream incision due to uplift or baselevel lowering. Without an understanding of process, serious errors of interpretation can be expected because, as in the case of hillslopes, convex, concave and straight slope segments can be formed in several ways (Table 4.2).

This discussion of process will be brief because it is probably sufficient to indicate that as far as is possible the physics and chemistry of a system should be understood. Of course, depending on the problem, additional sources of information as indicated on Fig. 1.1 could be required. In addition, information on rates of change will provide insight into the efficiency of the processes. Rates of change can be determined from historical studies as outlined above or from a model developed by LTS and, of course, by careful, long-term measurements in the field (Fig. 3.24).

Comparisons

Several of the problems (Table 4.1) can be approached through comparative studies by looking at a particular problem in the system context. For example, if one is asked to evaluate the stability of a site it is wise to search for similar site conditions within the same general area, and to use these to aid in the specific site evaluation.

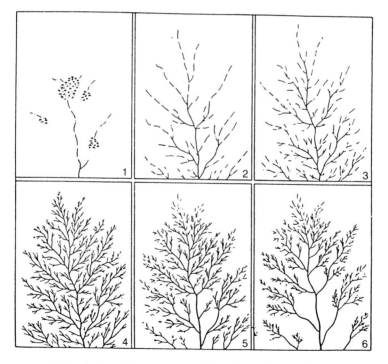

Figure 4.5. Evolution of drainage network based upon location for time substitution by Glock (1931). Each pattern was obtained from a different topographic map and arranged in a sequence from young (1) to old (6).

The Ohio River bank stability problem is a good example of the use of this technique (p. 58). By looking at bank conditions along the river the eroding sites were seen to be largely the natural consequence of river behavior instead of human influences. Comparative studies of this type can be used to evaluate the problems of space, location, sensitivity, singularity, complexity and multiplicity (Table 4.1).

This approach is similar to the location for time substitution, as described above, except that it is the present conditions rather than an evolutionary model that needs to be evaluated. This location for condition evaluation (LCE) has been used to identify sensitive alluvial valley floors (Fig. 3.26), a river's susceptibility to pattern change (Figs. 3.18, 4.6) and alluvial-fan stability (Fig. 3.27). Therefore, it is a means of identifying threshold conditions and the relative sensitivity of landforms (Schumm, 1988).

In each of these cases data were collected at a number of locations, and a quantitative relation was developed, that could lead to the identification of threshold conditions. For example, the slope of the line on Fig. 3.26 identifies a valley floor slope at a given drainage area at which gullies are likely to form. The curve of Fig. 3.18, when developed for a specific river can be used to identify when a river pattern is susceptible to change from meandering to braided and vice versa. When within an area a relation such as that of Fig. 3.26 is developed between drainage area and alluvial-fan slope, alluvial fans that are susceptible to fanhead trenching can be identified. The LCE approach, which involves collecting data for a number of similar landforms in an area, is a means of determining the condition or relative sensitivity of a single landform (Fig. 4.6).

Both the location for time substitutions (LTS) and the location for condition evaluation (LCE) involve the collection of data at a number of locations (Figs. 4.3, 4.6) and the utilization of the data to develop an evolutionary model (LTS) or to determine the sensitivity of a site (LCE). Both are valuable techniques that have been used primarily by geomorphologists for practical purposes of prediction, as well as for postdiction. Of even greater value is the fact that both techniques require that the investigator back away from a single site and look at many sites, which provides the 'big picture' and a basis for generalization.

Obviously in comparative studies a large sample that includes a large range of the independent variable is needed. For example, in Fig. 3.24 if there were only a few measurements an inverse relation could be a possibility, considering the scatter of the data. If the range of slope sampled was small, say between 0.1 and 0.3, no significant relation would have emerged. Therefore, comparative studies of similar features are desirable if premature conclusions, that are based upon limited observations, are to be avoided.

Finally, when dealing with a range of phenomena an understanding of one cannot be obtained by narrowly focusing on it to the exclusion of others. For example, I am convinced that an understanding of river meandering will not come from detailed studies of the hydraulics of flow and sediment transport through bends but from a comparison of the morphology and hydraulics of straight, meandering and braided channels. Therefore, an understanding of one landform type will result from an investigation of the spectrum of homologous landforms. In simplest terms the development of a useful explanatory classification of any phenomenon will not usually result from a narrowly focused study but rather from an integration of wider investigations.

If an understanding of complex phenomena is required it is frequently necessary to eliminate as many variables as possible and to select a simple situation. In the field it is often advisable to spend much time selecting a site or a location where a phenomenon can be studied under relatively simple conditions. This, of course, is the great advantage of experimental studies.

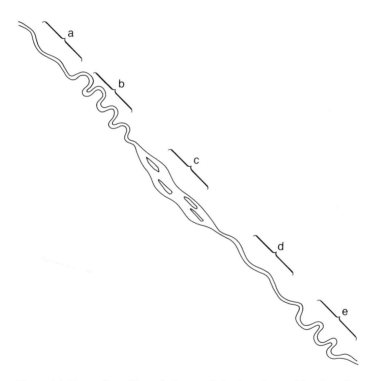

Figure 4.6. Data collected in each river reach (a–e) can be used for a location for condition evaluation (LCE). Development of a relation between sinuosity and valley slope (e.g. Fig. 3.18) will permit identification of sensitive channel reaches. For example, reaches b and c may both be close to the meandering–braided threshold of Fig. 3.18, and both may be likely to change pattern from meandering to braided and vice versa.

FINAL EXAMPLE

This final example is given in order to illustrate how some of the ten problems are considered and how the three general solutions are used in the investigation of a problem of some significance in the southwestern United States. As are most of the previous examples, this is a personal one that has been of interest to me since about 1962. It is a particularly good example because six working hypotheses were formulated during about twenty years, and the explanation–solution is undoubtedly composite.

In 1962 a report on drought in the southwestern United States was published in which it was concluded that there had been a dramatic and abrupt decrease in the sediment load in the Colorado River in the early 1940s (Fig. 4.7). Depending upon annual discharge, this decrease ranged from 50 to 100 million tons of sediment per year (Thomas, 1962).

As most of the water in the river is derived from the Rocky Mountains and as most of the sediment is derived from the Colorado Plateau, Thomas concluded that the decrease of sediment was the result of drought in the high sediment production areas of Utah and Arizona but above normal precipitation in the hardrock mountain areas of Wyoming and Colorado (Fig. 4.8). This explanation (Fig. 4.9, drought) appeared to be reasonable, although upon discussing it with US Geological Survey colleagues, I was informed that at about the same time the Water Resources Division had changed suspended-sediment samplers and that the sediment-load change reported by Thomas (Fig. 4.7) undoubtedly was the result of altered sampling techniques (Fig. 4.9, H_2, sampler).

The upper Colorado River above the Grand Canyon gaging station has a drainage area of 356 900 km^2, and it drains all of one and parts of four physiographic provinces: Colorado Plateau, Middle and Southern Rocky Mountains and a small part of the eastern Basin and Range. Without some expertise in this vast and complex area it was not possible in 1962 to evaluate the two hypotheses or even to generate new ones. However, in 1974 Hadley (1974, 1977) proposed that the decrease of sediment load could have been a result of a decrease of grazing pressures and increased erosion control efforts. For example, between 1941 and 1955 the number of sheep and goats in parts of the area were reduced by about 750 000. In addition, during the 1930s, thousands of small reservoirs and erosion control structures were built (Fig. 4.9, H_3, erosion control). As improbable as this hypothesis appeared at first, nevertheless, it conformed to the results of some experimental studies that showed abrupt changes of alluvial-fan morphology as a result of decreasing sediment delivery (Schumm et al., 1987, p. 314). It seemed probable, therefore, that something similar could have happened in the Colorado River basin with the effects of 20 years of conservation and flood control efforts suddenly appearing in the sediment-load records (Schumm, 1977, p. 325).

Of course, through the years detailed studies of channel morphology and alluvial stratigraphy were carried out by numerous investigators. For example, Hereford (1984, 1986), as a result of his studies of alluvial stratigraphy, was able to demonstrate that there was sediment deposition and floodplain formation in the lower reaches of the Little Colorado and the Paria Rivers during the 1940s to mid-1950s (Fig. 4.8), as a result of low peak flows and below normal precipitation (Fig. 4.9, H_4, hydrology). Certainly, elsewhere in western United States reduced peak discharge caused deposition and sediment storage (Schumm, 1977, pp. 159–64),

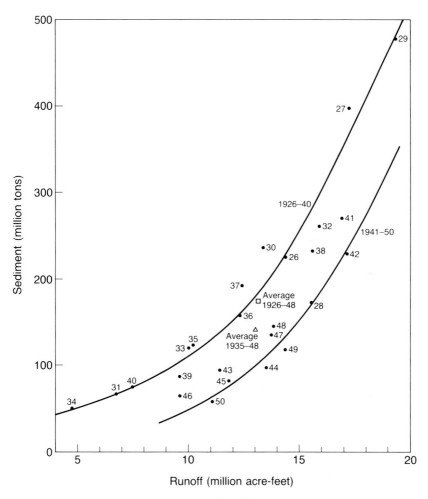

Figure 4.7. Relation between annual runoff and suspended-sediment load of the Colorado River in the Grand Canyon from 1926 to 1957 (from Thomas, 1962).

and on a regional basis this could have significantly reduced sediment delivery to the Colorado River.

Another factor was the invasion of salt cedar (tamarisk), which migrated upstream at a rate of about 20 km per year, into the Colorado River system. Although not cited as an explanation of the decreased sediment load, Graf (1978)

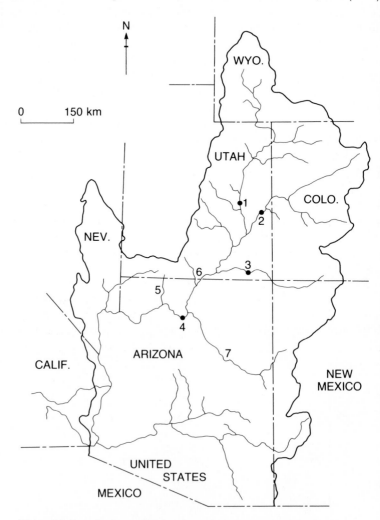

Figure 4.8. Location map of Colorado River basin. Numbers indicate locations of US Geological Survey gaging stations and tributaries as follows: 1. Green River near Green River, Utah, 2. Colorado River near Cisco, Utah, 3. San Juan River near Bluff, Utah, 4. Colorado River near Grand Canyon, 5. Kanab Creek, 6. Paria River, 7. Little Colorado River.

demonstrated that the appearance of tamarisk was associated with stabilization of bars and floodplains, which resulted in additional sediment storage. This could also have reduced Colorado River sediment loads (Fig. 4.9, H_5, vegetation).

Aggradation has been reported by several workers for a number of stream channels throughout the Colorado River Basin (Emmett, 1974; Leopold, 1976; Hereford, 1984, 1986; Graf, 1987). For example, in the Paria River basin, about 40 million m^3 of sediment accumulated in an area of 20 km^2 between about 1940 and 1980 (Hereford, 1986). In the same period of time in ten other drainage basins in the Colorado Plateau, that range in size from 10 to 1790 km^2, total sediment storage ranges from 235 million to 20 billion m^3 (Graf, 1987). Substantial changes in channel width and depth accompanied sediment storage and floodplain development. The Little Colorado River above its junction with Colorado River was a broad, braided stream in 1914; now the channel is confined in a well-vegetated floodplain, and it is only 50 per cent of its earlier width (Fig. 4.10). Abundant photographic evidence indicates clearly that sediment has accumulated in tributaries of the Colorado River (Graf, 1987; Hereford, 1984, 1986).

By about 1984 there were five hypotheses that could explain the decreased sediment loads in Colorado River, but because of the documented sediment storage in the tributary valleys a combination of H_4 and H_5 (Fig. 4.9) appeared as the most likely explanation of the reduced sediment load, and it should be noted that all of the hypotheses were developed in order to explain an abrupt decrease of sediment load in the early 1940s.

During the 20 years after 1962 my understanding of the channels, and the geology and geomorphic history of the Colorado River basin, had greatly improved, as a result of considerable field experience and background reading. Also a series of experimental studies provided information on the effects of base-level change on channel morphology and sediment production (Fig. 4.11), and field studies of incised streams that had been channelized in northern Mississippi provided a model of how channels in alluvium respond through time to incision (Fig. 4.4).

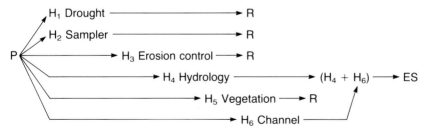

Figure 4.9. Development of composite explanation for decrease of sediment load in Colorado River.

With this background, when Graf (1985) plotted the sediment load data as a time series, the resemblance of this plot (Fig. 4.12) to the curves of Fig. 4.11 produced a new hypothesis, which is that the decrease of sediment load was the natural result of incised channel evolution (Fig. 4.4) since about 1880 (Fig. 4.9, H_6, channel), and that the change rather than being abrupt was a progressive decline.

The initiation of high sediment production and high sediment loads began in the late nineteenth century with arroyo cutting in many Colorado Plateau streams that drain the principal sediment-producing region of the Colorado River Basin. The literature on this topic is large, and the causes of regional stream entrenchment are still not well understood (Cooke and Reeves, 1976; Graf, 1983). Although beginning dates vary regionally (Webb, 1985; Graf, 1987), from 1865–1915 hundreds of arroyos were incised in several states, in many drainages, and in a wide variety of environments. Valley floors were deeply incised with devastating effects on the fragile agricultural economy of the region, and many pioneer settlements and farms were abandoned.

The extent of arroyo cutting in these valleys was substantial. In 1849, Chaco River in northwest New Mexico was 2.4 m wide and 0.5 m deep. In 1925, the river was 46–137 m wide and 6–9 m deep (Bryan, 1925). Early reports indicate that arroyos incised rapidly and produced large amounts of sediment; for example, in three years (1885–1888) Kanab Creek, Utah, formed a channel 18 m deep, 21 m wide, and 24 km long (Gregory, 1917). The channels in the Paria River basin of southern Utah were incised between 1883–1890 (Gregory and Moore, 1931).

As described above, through the years since 1962 when Thomas published his explanation, more information became available, and five additional hypotheses were advanced in order to describe the changed and changing sediment loads of the Colorado River. By the mid-1980s it seemed appropriate to test the six hypotheses, which required additional historical research, fieldwork involving location for time substitution, and comparisons among channels, in short, all of the general procedures outlined above (Gellis, 1988).

The drought hypothesis (H_1) was found to be improbable because sediment loads continued to decrease in the high sediment producing areas of the Colorado Plateau after the drought ended (Fig. 4.12), and the sampler hypothesis (H_2) was disproved because comparative tests revealed that the new sampler caught slightly more sediment than did the old one (Gellis, 1988).

The land-use hypothesis (H_3) was difficult to test except indirectly. Although locally erosion control measures would be effective, they probably did not influence sediment loads of the major rivers, and it is well known that removal of sediment from a channel causes replacement of the lost sediment by erosion of sediment previously stored in the valley. Therefore, small erosion control structures probably would not be very effective. In addition, many failed a short time after construction. In addition, Graf (1986) found that annual livestock numbers could not account for variation in sediment yield between 1930 and 1960. Never-

112

Final example

(*a*)

(*b*)

Figure 4.10. Photographs of Little Colorado River system (*a*) upstream in New Mexico where channel is unstable with much bank erosion and sediment production (Stage c, Fig. 4.4), (*b*) downstream in Arizona. The floodplain formed since 1940 (Stage e, Fig. 4.4).

113

theless, although some sediment load reduction could occur, as a result of erosion control, hypothesis H_3 was rejected as having a possible but only minor effect on sediment loads.

The combination of decreased flood peaks (H_4) and vegetation colonization of the valley floors (H_5) undoubtedly reduced sediment loads, as sediment was stored in the new floodplains. Therefore, H_4 is accepted as a likely explanation for some decrease of sediment loads beginning in the late 1940s.

The effect of vegetation (H_5), as a stabilizing influence on sedimentary deposits is well known, but generally sediment is deposited first and then the deposits are colonized by vegetation in these drylands. Therefore, although vegetation was an important stabilizing influence on the deposited sediment, it probably did not cause initiation of deposition, and H_5 can be rejected as not being an initial cause of sediment load reduction.

If as suggested the decrease of sediment loads commenced prior to 1940 then

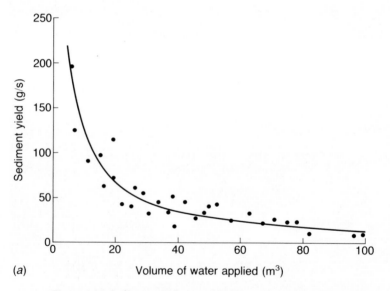

(a)

Figure 4.11. Sediment production during experimental studies of effect of baselevel lowering. (a) Sediment-yield variation observed during experimental drainage basin evolution (Parker, 1977; Schumm *et al.*, 1987), (b) Sediment yield during incised-channel experiment (Begin, 1979).

H_6 appears to be the most probable explanation. Numerous studies have demonstrated that following channel incision, sediment loads rise to a maximum and then decline as the channel evolves to a new condition of relative stability (Figs. 3.5, 4.11). Arroyos in the Colorado Plateau are very similar to incised channels formed by channelization, and the model of incised-channel evolution developed by location for time substitution (LTS) (Fig. 4.4) is valid for arroyos. Field work in the Colorado Plateau documented similar channel change along the arroyos, as based upon LTS (Gellis, 1988). Downstream, where the channels have widened and floodplains have formed, sediment is trapped and stored (Fig. 4.4(*d*), (*e*), Fig. 4.10(*b*)). Upstream channel widening and sediment production continues (Fig. 4.4(*b*), (*c*), Fig. 4.10(*a*)). The similarity between morphology and evolution of the arroyos and the channelized streams and the change of sediment load to be expected during this evolution (Fig. 4.11) suggests that the natural evolution of incised channels will cause a progressive rather than abrupt change of sediment

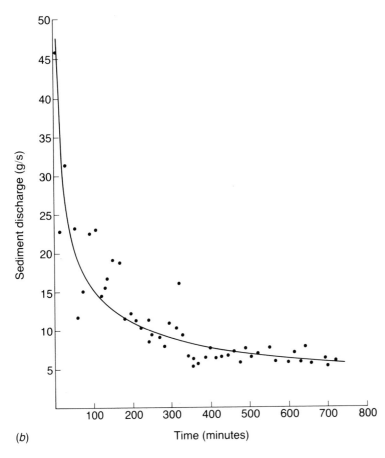

(*b*)

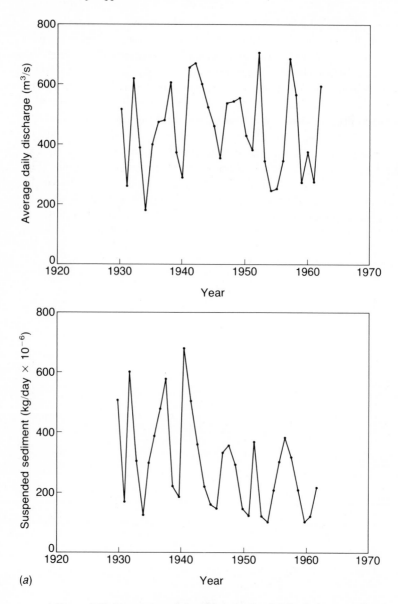

Figure 4.12. Mean suspended sediment load and daily discharge, Colorado River (*a*) and Plateau Country (*b*). Data for Plateau County, which is the area of major arroyo incision, was obtained by subtracting the sediment loads of the Green River at Green River, Utah, the Colorado River near Cisco, Utah, and

116

Final example

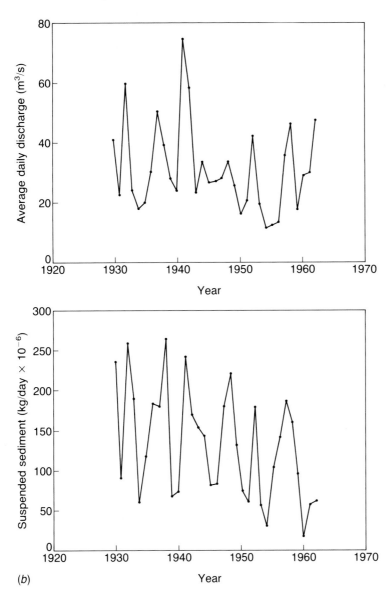

(b)

the San Juan River near Bluff, Utah from the sediment load of the Colorado River near Grand Canyon (Fig. 4.8). Records after 1962 cannot be used because of the closure of large dams upstream (Glen Canyon, Flaming Gorve, Navajo).

load. Therefore, H_6 is a very probable explanation of the major changes shown on Fig. 4.12 (Gellis *et al.*, 1990; Schumm and Gellis, 1989).

Nevertheless, the explanation of the changes of sediment load in Colorado River must be a composite explanation. As noted above, two of the hypotheses seem to be the most probable explanations of this change. H_6 explains the overall 30-year change (Fig. 4.12) but H_4 explains the documented abrupt decrease in the early 1940s as described by Thomas (1962). Therefore, superimposed on a declining trend is an abrupt change in the early 1940s that was produced by a hydrologic fluctuation. The decreased floodpeaks (H_4) also permitted colonization of the alluvial deposits by vegetation to form the floodplains that were so obvious after the decade of the 1940s.

In this example of the development of a probable solution to a complex problem, many of the ten problems were involved, and the three general problem-solving approaches were used. For example, time was a problem because Thomas (1962) only had 20 years of record, but the events of the late nineteenth century required a record of at least 80 years. Even today only 50 per cent of the needed record is available, but in order to explain the change one had to think beyond the short record. Size was a problem because the initial advancement of H_6 was based on experimental results from a 150 m^2 experimental facility that were applied to the 356 900 km^2 upper Colorado River basin (Fig. 4.11(*a*)).

Location was a problem because some studies were restricted to the lower parts of tributary basins and observations made in incised channels of northern Mississippi were used to develop the incised-channel evolution model that was applied to the arroyos (Fig. 4.4). Indeed it was only during a reconnaissance flight up the Little Colorado River and the arroyos of the Navajo Indian Reservation that it was recognized that, as one advanced upstream, the channel changed from a stage e to a stage b channel of Fig. 4.4 (Fig. 4.10).

The advancement of six different hypotheses is a clear indication that convergence was a significant problem, but multiplicity was not so much a problem as a solution (Fig. 4.9). Singularity was a problem because initially the question could be raised that the changes noted by Thomas were simply scatter of the sediment load data (Fig. 4.7). Although complex behavior was observed in these channels, complexity did not become a problem nor did divergence, efficiency or sensitivity.

The solution or partial solution of the problem involved history, because until the erosional development of the arroyos of the late nineteenth century was appreciated the assumption of very high sediment loads in the late nineteenth and early twentieth centuries could not have been assumed.

An understanding of processes of channel incision, as outlined in Fig. 4.4 and the changes of sediment delivery (Fig. 3.5) plus the results of experimental studies (Fig. 4.11), were needed before hypothesis 6 could be advanced.

Therefore, as stated above, an understanding of process is essential to the

118

solution of any earth science problem. Finally, the comparative field work among many channels of the Colorado Plateau, as well as channels of northern Mississippi, provided the basis for the location for time substitution (Fig. 4.4) that produced the composite explanation of Fig. 4.9.

It should not be assumed that the attempt to explain Colorado River sediment load changes was strictly an academic exercise. An understanding of this great fluvial system has great significance for the life of reservoirs and for erosion control in this vast dryland region. Therefore, future work will undoubtedly produce other hypotheses, and the reader is warned that many workers in the Colorado Plateau country will not accept H_6 as the solution to this question of sediment load change.

FINAL WORDS

The discussion of ten problems faced during explanation and extrapolation is not an attempt to discourage scientific investigation. Rather, it is an attempt to emphasize the complexity of natural systems and to explain why explanation and extrapolation can often be suspect. The identification of these problems does not by a process of name magic solve anything, but it does help to develop a rational scientific approach to complex systems. Furthermore, recognition of the problems will lead to more thorough research plans. Consideration of the problems may, therefore, be time-consuming, difficult and expensive, but never as expensive as failure. The development of the required understanding of natural systems will be intellectually rewarding, and it will be cost effective.

A scientific approach rather than a specific scientific method is the basis for research accomplishment. At least for geomorphic investigations both the location for time substitution (LTS) and the location for condition evaluation (LCE) provide significant assistance in the development of evolutionary models and the identification of sensitive landforms.

It is difficult to challenge Beveridge's (1957) conclusion that the best way to learn how to do science is to go do it, and this is especially true where field observations are required, as in the earth sciences. Nevertheless, I leave you with a strong recommendation to read Gilbert (1886, 1896), Chamberlin (1890), Beveridge (1957), Medawar (1979a, 1984), Parker (1986), Strahler (1987) and Lewis Thomas (1974, 1979, 1983) because these scientists attempted to provide guidelines for research. Finally as Luna B. Leopold advised me many years ago, 'Always ask yourself important questions'.

References

Adams, J. (1980). Active tilting of the United States midcontinent: Geodetic and geomorphic evidence. *Geology*, **8**, 442–6.

Ager, D. G. (1981). *The nature of the stratigraphic record*, 2nd edn. New York: Halsted Press. 122 pp.

Albritton, C. C. (ed.) (1963). *The fabric of geology*. Reading, MS: Addison–Wesley. 372 pp.

Albritton, C. C. (1967). Uniformity and simplicity. *Geol. Soc. Am., Spec. Pap.*, **89**, 99 pp.

Anderson, R. Y. (1986). The varve microcosm: Propagation of cyclic bedding. *Paleoceanography*, **1**, 373–82.

Arnett, R. R. (1979). The use of differing scales to identify factors controlling denudation rates. In *Geographical approaches to fluvial processes*, ed. A. F. Pitty, pp. 127–47. Norwich, UK: Geobooks. 300 pp.

Baker, V. R. (1973). Paleohydrology and sedimentology of Lake Missoula flooding in eastern Washington. *Geol. Soc. Am., Spec. Pap.*, **144**, 79 pp.

Baker, V. R. (1977). Stream-channel response to floods, with examples from central Texas. *Geol. Soc. Am. Bull.*, **88**, 1057–71.

Baker, V. R. and Milton, D. J. (1974). Erosion by catastrophic floods on Mars and Earth. *Icarus*, **23**, 27–41.

Barber, B. (1961). Resistance by scientists to scientific discovery. *Science*, **134**, 596–602.

Begin, Z. B. (1979). Aspects of degradation of alluvial streams in response to base-level lowering. Unpublished PhD dissertation, Colorado State University, Fort Collins, CO. 239 pp.

Begin, Z. B. and Schumm, S. A. (1979). Instability of alluvial valley floors: A method for its assessment. *Am. Soc. Agr. Eng.*, **22**, 347–50.

Begin, Z. B. and Schumm, S. A. (1984). Gradational thresholds and landform singularity: Significance for quaternary studies. *Quaternary Research*, **31**, 267–74.

Bergstrom, F. W. and Schumm, S. A. (1981). Episodic behavior in badlands. *Internat. Assoc. Sci. Hydrol. Pub.*, **132**, 478–89.

Beven, K. (1981). The effect of ordering on the geomorphic effectiveness of hydrologic events. *Internat. Assoc. Sci. Hydrol. Publ.*, **132**, 510–26.

Beveridge, W. I. B. (1957). *The art of scientific investigation*, 3rd edn. New York: Norton.

Beveridge, W. I. B. (1980). *The seeds of discovery*. New York: Norton. 130 pp.

Bibby, C. (1960). *T. H. Huxley, scientist, humanist and education*. New York: Horizon Press. 330 pp.

Blackmore, S. (1986). *The adventures of a parapsychologist*. Buffalo, New York: Prometheus Books. 249 pp.

Boison, P. J. and Patton, P. C. (1985). Sediment storage and terrace formation in Coyote Gulch basin, south-central Utah. *Geology*, **13**, 31–4.

Boorstin, D. J. (1983). *The Discoverers*. New York: Random House. 745 pp.

Born, S. M. and Ritter, D. F. (1970). Modern terrace development near Pyramid Lake, Nevada and its geologic implications. *Geol. Soc. Am. Bull.*, **81**, 1233–42.

Bowen, N. L. (1948). The granite problem and the method of multiple prejudices. *Geol. Soc. Am., Mem.*, **28**, 79–90.

Boyce, R. C. (1975). Sediment routing with sediment delivery ratios. *Agricultural Research Service*, ARS-S-40, pp. 61–5.

Bretz, J. H. (1969). The Lake Missoula floods and the channeled scabland. *J. Geol.*, **77**, 505–43.

Brunet, R. (1968). *Les phenomenes de discontinuite en geographie*. Memoires et Documents, Editions du Centre National Recherche Scientifique, vol. 7, 117 pp.

Brunsden, D. and Thornes, J. B. (1979). Landscape sensitivity and change. *Inst. Brit. Geogr., Trans.*, **4**, 468–84.

Bryan, K. (1925). Date of channel trenching (arroyo cutting) in the arid southwest. *Science*, **62**, 338–44.

Bubnoff, S. Von (1963). *Fundamentals of geology*. Edinburgh: Oliver and Boyd. 286 pp.

Bucher, W. H. (1936). The concept of natural law in geology. *Science*, **84**, 491–8.

Bucher, W. H. (1941). The nature of geological inquiry and the training required for it. *Am. Inst. Mining Met. Eng., Tech. Pub.*, **1377**, 1–6.

Bunge, M. (1984). What is pseudo science? *The Skeptical Inquirer*, **9**, 36–46.

Burchfield, J. D. (1975). *Lord Kelvin and the age of the earth*. New York: Science History Publications. 260 pp.

Burnett, A. W. and Schumm, S. A. (1983). Neotectonics and alluvial river response. *Science*, **222**, 49–50.

Carson, M. A. (1971). Application of the concept of threshold slopes to the Laramie Mountains, Wyoming. *Inst. Brit. Geogr. Spec. Publ.*, **3**, 31–47.

Challinor, J. (1968). Uniformitarianism – the fundamental principles of geology. *23rd Internat. Geol. Congr. (Prague)*, **13**, 331–47.

Chamberlin, T. C. (1890). The method of multiple working hypotheses. *Science*, **15**, 92–6. (Reprinted 1965, *Science*, **148**, 754–9.)

Chamberlin, T. C. (1897). The method of multiple working hypotheses. *J. Geol.*, **5**, 837–48. (Reprinted 1931, *J. Geol.*, **39**, 155–65. Reprinted 1944, *Scientific Monthly*, **59**, 356–62.)

Chargaff, E. (1978). *Heraclitean fire*. New York: Rockefeller University Press. 252 pp.

Chorley, R. J. (1962). *Geomorphology and general systems theory*. US Geological Survey Professional Paper 500-B, 10 pp.

Chorley, R. J. and Kennedy, B. A. (1971). *Physical geography: A systems approach*. New York: Prentice-Hall. 370 pp.

Church, M. (1980). Records of recent geomorphological events. In *Timescales in geomorphology*, ed. R. A. Cullingford, D. A. Davidson and J. Lewin, pp. 13–29. Chichester: John Wiley & Sons.

Church, M. and Slaymaker, O. (1989). Disequilibrium of Holocene sediment yield in glaciated British Columbia. *Nature*, **337**, 452–4.

References

Clark, J. A., Farrell, W. E. and Peltier, W. R. (1978). Global changes in postglacial sea level: A numerical calculation. *Quaternary Res.*, **9**, 265–87.

Clark, M. J. (1987). The alpine sediment system: A context for glacio-fluvial processes. In *Glacio-fluvial sediment transfer*, ed. A. M. Gurnell and M. J. Clark, pp. 9–32. New York: Wiley.

Cooke, R. U. and Reeves, R. W. (1976). *Arroyos and environmental change in the American Southwest*. Oxford, England: Clarendon Press. 213 pp.

Cooper, H. F., Jr (1977). A summary of explosion cratering phenomenon relevant to meteor impact events. In *Impact and explosion cratering*, ed. D. J. Roddy, R. O. Pepin and R. B. Therrill, pp. 11–44. New York: Pergamon Press.

Coyne, A. M. (1984). *Introduction to inductive reasoning*. Lanham, MD: University Press of America. 282 pp.

Crittenden, M. D., Jr (1963). *New data on the isostatic deformation of Lake Bonneville*. US Geological Survey Professional Paper 454-E, 31 pp.

Darwin, C. G. (1956). The time scale in human affairs. In *Man's role in changing the face of the earth*, ed. W. L. Thomas Jr., Chicago: University of Chicago Press, pp. 963–9.

Davis, W. M. (1902). Baselevel, grade and peneplain. *J. Geol.*, **10**, 77–111.

Delcourt, H. R., Delcourt, P. A. and Webb, T. III (1983). Dynamic plant ecology: The spectrum of vegetational change in space and time. *Quaternary Science Revs.*, **1**, 153–75.

Dott, R. H., Jr (1983). Episodic sedimentation. *J. Sediment. Petrol.*, **53**, 5–23.

Douglas, I. (1980). Climatic geomorphology: Present-day processes and landform evolution, problems of interpretation. *Zeit. Geomorph. Supplementband*, **36**, 24–47.

Emmett, W. W. (1974). Channel aggradation in western United States as indicated by observations at Vigil Network Sites. *Zeit. Geomorph. Supplementband*, **21**, 52–62.

Erskine, W. D. and Warner, R. F. (1988). Geomorphic effects of alternating flood- and drought-dominated regimes on NSW coastal river. In *Fluvial geomorphology Australia*, ed. R. F. Warner, pp. 223–44. Sydney: Academic Press.

Feibleman, J. K. (1972). *Scientific method*. The Hague: Martinus Nijhoff. 246 pp.

Feigl, H. (1953). The scientific outlook: Naturalism and humanism. In *Readings in the philosophy of science*, ed. Herbert Feigl and May Brodbeck, pp. 8–18. New York: Appleton-Century Crofts.

Feyerabend, P. (1975). *Against method*. London: New Left Books. 339 pp.

Feyerabend, P. (1978). *Science in a free society*. London: New Left Books. 221 pp.

Fisk, H. N. (1947). *Fine-grained alluvial deposits and their effects on Mississippi River activity*. Vicksburg, US Army Engineers Waterways Experiment Station, vol. 2.

Ford, D. (1988). Crown of Thorns. *The New Yorker*, **July 25**, 34–63.

Franklin, H. (1980). Hess's development of his sea floor spreading hypothesis. In *Scientific discovery: Case studies*, ed. T. Nickels, pp. 345–66. Boston: Reidel.

Froehlich, W., Kazowski, L. and Starkel, L. (1977). Studies of present-day and past river activity in the Polish Carpathians. In *River channel changes*, ed. G. Gregory, pp. 411–28. New York: Wiley Interscience.

Gage, M. (1970). The tempo of geomorphic change. *J. Geol.*, **78**, 619–25.

Gardner, T. W., Jorgensen, D. W., Shuman, C. and Lemieux, C. R. (1987). Geomorphic and tectonic process rates: Effects of measured time intervals. *Geology*, **15**, 259–61.

References

Gary, M., McAfee, R., Jr and Wolf, C. L. (1972). *Glossary of geology*. Washington, DC: American Geological Institute. 805 pp.

Geike, A. (1905). *Founders of geology*, 2nd edn. London: Macmillan. 486 pp.

Gellis, A. C. (1988). Decreasing sediment and salt loads in the Colorado River basin: A response to arroyo evolution. Unpublished M. S. thesis, Colorado State University, Fort Collins, CO, 178 pp.

Gellis, A. C., Hereford, R., Schumm, S. A. and Hayes, B. (1990). Channel evolution and hydrologic variations in the Colorado River Basin: Factors influencing sediment and salt loads. *J. Hydrol.* (in press).

Giere, R. N. (1988). *Explaining science*. Chicago: University of Chicago Press. 321 pp.

Gilbert, G. K. (1886). The inculcation of the scientific method by example with an illustration drawn from the Quaternary geology of Utah. *Am. J. Sci.*, **31**, 284–99.

Gilbert, G. K. (1896). The origin of hypotheses, illustrated by the discussion of a topographic problem. *Science*, **3**, 1–12.

Gilbert, G. K. (1904). A case of plagiarism. *Science*, **20**, 115–16.

Gingerich, P. D. (1976). Paleontology and phylogeny: patterns of evolution at the species level in Tertiary mammals. *Am. J. Sci.*, **276**, 1–28.

Gleason, H. A. (1926). The individualistic concept of the plant association. *American Midland Naturalist*, **21**, 92–100.

Glock, W. S. (1931). The development of drainage systems: A synoptic view. *Geogr. Rev.*, **21**, 475–82.

Gould, S. J. (1965). Is uniformitarianism necessary? *Am. J. Sci.*, **263**, 223–8.

Gould, S. J. (1984). Toward the vindication of punctuational change. In *Catastrophes and earth history*, ed. W. A. Berggren and J. A. Van Couvering, pp. 9–34. Princeton: Princeton University Press.

Graf, W. L. (1978). Fluvial adjustments to the spread of tamarisk in the Colorado Plateau region. *Geol. Soc. Am. Bull.*, **89**, 1491–501.

Graf, W. L. (1982). Spatial variation of fluvial processes in semi-arid lands. In *Space and time in geomorphology*, ed. C. E. Thorne, pp. 193–217. London: George Allen and Unwin.

Graf, W. L. (1983). The arroyo problem: Paleohydrology and paleohydraulics in the short term. In *Background to paleohydrology*, ed. K. J. Gregory, pp. 279–302. London: John Wiley and Sons.

Graf, W. L. (1985). *The Colorado River, instability and basin management*. Washington, DC: Association American Geographers. 88 pp.

Graf, W. L. (1986). Fluvial erosion and federal public policy in the Navajo Nation. *Phys. Geogr.*, **7**, 97–115.

Graf, W. L. (1987). Late Holocene sediment storage in canyons of the Colorado Plateau. *Geol. Soc. Am. Bull.*, **99**, 261–71.

Gregory, H. E. (1917). *Geology of the Navajo Country*. US Geological Survey Professional Paper 93, 161 pp.

Gregory, H. E. and Moore, R. C. (1931). *Geology of the Kaiparowits region. A geographic and geologic reconnaissance of parts of Utah and Arizona*. US Geological Survey Professional Paper 220, 200 pp.

Gregory, K. J. (1985). *The nature of physical geography*. London: Edward Arnold.

Gregory, K. J. and Gardiner, V. (1975). Drainage density and climate. *Zeit. Geomorph.*, **19**, 287–98.

Gretener, P. E. (1967). Significance of the rare event in geology. *Am. Assoc. Petrol. Geol. Bull.*, **51**, 2197–206.

Gretener, P. E. (1984). Reflections on the rare event and related concepts in geology. In *Catastrophes in earth history*, ed. W. A. Berggren and J. A. Van Couvering, pp. 77–89. Princeton: Princeton University Press.

Griggs, G. B. (1987). The production, transport and delivery of coarse-grained sediment by California coastal streams. In *Coastal sediments '87*, ed. N. C. Kraus, pp. 1825–38. New York: American Society of Civil Engineers.

Grinnell, F. (1987). *The scientific attitude*. Boulder, CO: Westview Press. 141 pp.

Hadley, R. F. (1974). Sediment yield and land use in southwest United States. *Internat. Assoc. Sci. Hydrol. Publ.*, **113**, 96–8.

Hadley, R. F. (1977). Evaluation of land-use and land-treatment practices in semiarid western United States. *Phil. Trans. R. Soc. London*, **278**, 543–54.

Hagner, A. F. (1963). Philosophical aspects of the geological sciences. In *The fabric of geology*, ed. C. C. Albritton, pp. 233–41. Reading: Addison Wesley.

Haines-Young, R. H. and Petch, J. R. (1983). Multiple working hypotheses: Equifinality and the study of landforms. *Inst. Brit. Geogr. Trans.*, **8**, 458–66.

Haines-Young, R. H. and Petch, J. R. (1986). *Physical geography: Its nature and methods*. New York: Harper and Row. 230 pp.

Hallam, A. (1983). *Great geological controversies*. Oxford: Oxford University Press. 182 pp.

Hallam, A. (1987). End-cretaceous mass extinction event: Argument for terrestrial causation. *Science*, **238**, 1237–42.

Harré, R. (1960). *An introduction to the logic of the sciences*. London: Macmillan. 180 pp.

Hartshorn, J. H. (1967). Geology of the Taunton Quadrangle, Bristol and Plymouth Counties, Massachusetts. *US Geol. Surv. Bull.*, **1163-D**, 67 pp.

Harvey, A. M. and Bordley, J. III (1970). *Differential diagnosis*. Philadelphia: W. B. Saunders. 1238 pp.

Harvey, D. (1969). *Explanation in geography*. London: Edward Arnold. 521 pp.

Hereford, R. (1984). Climate and ephemeral-stream processes: Twentieth-century geomorphology and alluvial stratigraphy of the Little Colorado River, Arizona. *Geol. Soc. Am. Bull.*, **95**, 654–68.

Hereford, R. (1986). Modern alluvial history of the Paria River drainage basin, southern Utah. *Quaternary Res.*, **25**, 293–311.

Hey, R. D. (1979). Dynamic process-response model of river channel development. *Earth Surface Proc.*, **4**, 59–72.

Hsü, K. J. (1983). *The Mediterranean was a desert*. Princeton: Princeton University Press. 189 pp.

Huggett, R. J. (1988). Terrestrial catastrophism: Causes and effects. *Prog. Phys. Geogr.*, **12**, 509–32.

Hupp, C. R., Osterkamp, W. R. and Thornton, J. L. (1987). *Dendrogeomorphic evidence and dating of recent debris flows on Mount Shasta, northern California*. US Geological Survey Professional Paper 1396-B, 39 pp.

Johnson, D. (1933). Role of analysis in scientific investigation. *Geol. Soc. Am. Bull.*, **44**, 461–94.

Johnson, D. (1940). Mysterious craters of the Carolina Coast: A study in methods of research. *Science in Progress*, 78–106.

Jong, W. J. (1966). Actualism in geology and geography. *Nederlandsch Aardrijkskundig Genootschap, Tijdschrist*, **83**, 238–48.

Judson, S. (1947). Large-scale superficial structures: A discussion. *J. Geol.*, **54**, 168–75.

Keller, G. (1989). Extended period of extinctions across the Cretaceous–Tertiary boundary in planktonic foraminifera of continental-shelf sections: Implications for impact and volcanism theories. *Geol. Soc. Am. Bull.*, **101**, 1408–19.

Kennedy, J. F. and Brooks, N. H. (1965). Laboratory study of an alluvial stream at constant discharge. *US Dept. Agr. Misc. Publ.*, **970**, 320–30.

Kidson, C. and Carr, A. P. (1959). The movement of shingle over the sea bed close inshore. *Geogr. J.*, **125**, 380–9.

Kilinc, M. and Richardson, E. V. (1973). *Mechanics of soil erosion from overland flow generated by simulated rainfall.* Colorado State University, Hydrology, Paper 63, 54 pp.

King, P. B. and Schumm, S. A. (1980). *The physical geography (geomorphology) of William Morris Davis.* Norwich, UK: Geobooks. 217 pp.

Kirkby, M. J. (1973). Landslides and weathering rates. *Geol. appl. e Idrageologia*, **8**, 171–83.

Kirkby, M. J. (1987). The Hurst Effect and its implications for extrapolating process rates. *Earth Surface Processes and Landforms*, **12**, 57–67.

Kitcher, P. (1982). *Abusing science: The case against creationism.* Cambridge, MA: MIT Press. 213 pp.

Kitts, D. B. (1963a). The theory of geology. In Albritton (1963), pp. 49–68.

Kitts, D. B. (1963b). Historical explanation in geology. *J. Geol.*, **71**, 297–313.

Kitts, D. B. (1966). Geologic time. *J. Geol.*, **79**, 127–46.

Kitts, D. B. (1973). Grove Karl Gilbert and the concept of 'hypothesis' in late ninteeenth century geology. In *Foundations of scientific method: The nineteenth century*, ed. R. N. Giere and R. S. Westfall, pp. 259–73. Bloomington: Indiana University Press.

Kitts, D. B. (1974). Physical theory and geological knowledge. *J. Geol.*, **82**, 1–23.

Kitts, D. B. (1977). *The structure of geology.* Dallas: Southern Methodist University Press. 180 pp.

Klemés, V. (1983). Conceptualization and scale in hydrology. *J. Hydrol.*, **65**, 1–23.

Klemés, V. (1985). Dilettantism in hydrology: Transition or destiny. *Water Resour. Res.*, **22**, 1775–885.

Knox, J. C. (1977). Human impacts on Wisconsin stream channels. *Ann. Assoc. Am. Geogr.*, **67**, 323–42.

Knox, J. C. (1983). Responses of river systems to Holocene climates. In *Late Quaternary environments of the United States, The Holocene*, ed. H. E. Wright, Jr, pp. 26–41. Minneapolis: University of Minnesota Press.

Koestler, A. (1978). *Janus.* New York: Random House. 354 pp.

Komar, P. D. (1983). Rhythmic shoreline features and their origins. In *Mega-geomorphology*, ed. R. Gardner and H. Scoging, pp. 92–112. Oxford: Oxford University Press.

Kozarski, S. and Rotnicki, K. (1977). Valley floors and changes of river channel patterns in the north Polish Plain during the late Wurm and Holocene. *Quaestiones Geographicae*, **4**, 51–93.

Kuhn, T. S. (1970). *The structure of scientific revolutions*, 2nd edn. Chicago: University of Chicago Press. 210 pp.

Kyte, F. T., Zhou, L. and Wasson, J. T. (1988). New evidence on the size and possible effects of a late Pliocene oceanic asteroid impact. *Science*, **241**, 63–5.

Lamb, H. H. (1977). *Climate, present, past and future*, vol. 2. London: Methuen. 835 pp.

Langbein, W. B. and Schumm, S. A. (1958). Yield of sediment in relation to mean annual precipitation. *Am. Geophys. Union Trans.*, **39**, 1076–84.

Laudan, R. (1980). The method of multiple working hypotheses and the development of plate tectonic theory. In *Scientific discovery: Case studies*, ed. T. Nickles, pp. 331-43. Boston: D. Reidel.

Leatherdale, W. H. (1974). *The role of analogy model and metaphor in science*. New York: American Elsevier. 276 pp.

Leopold, L. B. (1951). Rainfall frequency, an aspect of climatic variation. *Am. Geophys. Union Trans.*, **32**, 347–57.

Leopold, L. B. (1976). Reversal of erosion cycle and climatic change. *Quaternary Res.*, **6**, 557–62.

Leopold, L. B. and Langbein, W. B. (1962). *The concept of entropy in landscape evolution*. US Geological Survey Professional Paper 500-A, 20 pp.

Leopold, L. B. and Langbein, W. B. (1963). Association and indeterminacy in geomorphology. In Albritton (1967), pp. 184–92.

Leopold, L. B. and Wolman, M. G. (1957). *River channel patterns: braided, meandering and straight*. US Geological Survey Professional Paper 282-B, pp. 39–84.

Little, W., Fowler, H. W. and Coulson, J. (1964). *The Oxford universal dictionary*. London: Oxford University Press. 2515 pp.

Livingstone, D. N. and Harrison, R. T. (1981). Meaning through metaphor: Analog as epistemology. *Assoc. Am. Geogr., Ann.*, **71**, 95–107.

McCaulay, D. (1979). *Motel of the mysteries*. Boston: Houghton Mifflin Co. 96 pp.

McDowell, P. F. (1983). Evidence of stream response to Holocene climate change in a small Wisconsin watershed. *Quaternary Res.*, **19**, 100–16.

McKelvey, V. E. (1963). Geology as a study of complex natural experiments. In Albritton (1963), pp. 69–74.

McShea, D. W. and Raup, D. M. (1986). Completeness of the geologic record. *J. Geol.*, **94**, 569–74.

Mackay, A. L. and Ebison, M. (1977). *Scientific quotations: The harvest of a quiet eye*. New York: Crane, Russak and Co. 192 pp.

Mackin, J. H. (1948). Concept of the graded river. *Geol. Soc. Am. Bull.*, **59**, 463–511.

Mackin, J. H. (1963). Rational and empirical methods of investigation in geology. In Albritton (1963), pp. 135–63.

Mann, C. J. (1970). Randomness in nature. *Geol. Soc. Am. Bull.*, **81**, 95–104.

May, R. M. (1977). Thresholds and breakpoints in ecosystems with a multiplicity of stable states. *Nature*, **269**, 471–7.

Medawar, P. B. (1979a). *Advice to a young scientist*. New York: Harper and Row. 109 pp.

Medawar, P. B. (1979b). Foreword. In *The logic of scientific inference*, by Jennifer Trusted. London: Macmillan. 145 pp.

Medawar, P. B. (1984). *Pluto's republic*. Oxford: Oxford University Press. 351 pp.

Medvedev, Z. A. (1969). *The rise and fall of T. D. Lysenko* (translated by I. M. Lerner). New York: Columbia University Press. 284 pp.

Mitchell, J. M., Jr (1976). An overview of climatic variability and its causal mechanisms. *Quaternary Res.*, **6**, 481–93.

Morgan, R. P. C. (1973). The influence of scale in climatic geomorphology: A case study of drainage density in West Malaysia. *Geogr. Annal.*, **55A**, 107–15.

Morris, R. (1985). *Times arrows*. New York: Simon and Schuster. 240 pp.

Morris, W. (ed.) (1981). *The American heritage dictionary*. Boston: Houghton Mifflin. 1550 pp.

Mörner, N. A. (1984). Eustacy geoid changes, and multiple geophysical interaction. In *Catastrophes and earth history*, ed. W. A. Berggren and J. A. Van Couvering, pp. 395–415. Princeton: Princeton University Press.

Mosley, M. P. and Zimpfer, G. L. (1976). Explanation in geomorphology. *Zeit. Geomorph*, **20**, 381–90.

Muhs, D. R. (1984). Intrinsic thresholds in soil systems. *Physical Geography*, **5**, 99–110.

Mycielska-Dowgiallo, E. (1977). Channel pattern changes during the last glaciation and Holocene, in the northern part of the Sandomirez basin and the middle part of the Vistula valley, Poland. In *River channel changes*, ed. K. G. Gregory, pp. 75–87. New York: Wiley.

Nagel, E. (1961). *The structure of science*. New York: Harcourt, Brace and World. 618 pp.

Nairn, A. E. M. (1965). Uniformitarianism and environment. *Palaeogeogr., Palaeoclimatol., Palaeoecol.*, **1**, 5–11.

Paine, A. D. M. (1985). Ergodic reasoning in geomorphology: A preliminary review. *Prog. Phys. Geogr.*, **9**, 1–15.

Pantin, C. F. A. (1968). *The relations between the sciences*. Cambridge: Cambridge University Press. 206 pp.

Parker, R. B. (1986). *The tenth muse*. New York: Scribners. 221 pp.

Parker, R. S. (1977). Experimental study of basin evolution and its hydrologic implications. Unpublished Ph.D. dissertation, Colorado State University, Fort Collins, CO, 331 pp.

Patton, P. C. and Schumm, S. A. (1975). Gully erosion, Northern Colorado: A threshold phenomenon. *Geology*, **3**, 88–90.

Penning-Roswell, E. and Townshend, J. R. G. (1978). The influence of scale on the factors affecting stream channel slope. *Trans. Inst. Brit. Geogr.*, **3**, 395–415.

Phillips, J. D. (1988). The role of spatial scale in geomorphic systems. *Geographical Analysis*, **20**, 308–17.

Pitty, A. F. (1982). *The nature of geomorphology*. London: Methuen. 161 pp.

Platt, J. R. (1964). Strong inference. *Science*, **146**, 347–53.

Plotnick, R. E. (1986). A fractal model for the distribution of stratigraphic hiatuses. *J. Geol.*, **94**, 885–90.

Popper, K. R. (1968). *The logic of scientific discovery*, 2nd edn. New York: Harper and Row. 480 pp.

References

Pretorius, D. A. (1973). *The role of EGRU in mineral exploration in South Africa.* Economic Geology Research Unit, University of Witwatersrand Information Circular 77, 16 pp.

Pyne, S. J. (1978). Methodologies for geology: G. K. Gilbert and T. C. Chamberlin. *Isis,* **69,** 413–24.

Pyne, S. J. (1980). *Grove Karl Gilbert. A great engine of research.* Austin: University of Texas Press. 306 pp.

Rogers, R. D. (1989). Influence of sparse vegetation cover on erosion and rill patterns: An experimental study. Unpublished MS thesis, Colorado State University, Fort Collins, CO, 66 pp.

Rouché, B. (1984). *The medical detectives,* vol. 2. New York: E. P. Dutton. 367 pp.

Russell, B. (1961). *Religion and science.* Oxford: Oxford University Press. 256 pp.

Rutten, M. G. (1971). *The origin of life by natural causes.* Amsterdam: Elsevier. 420 pp.

Sacks, O. (1987). *The man who mistook his wife for a hat.* New York: Harper and Row. 243 pp.

Sadler, P. M. (1981). Sediment accumulation rates and the completeness of stratigraphic sections. *J. Geol.,* **89,** 569–84.

Sayer, A. (1984). *Method in social science.* London: Hutchinson and Co. 271 pp.

Schultz, P. H. (1976). *Moon morphology.* Austin: University of Texas Press. 626 pp.

Schumm, S. A. (1964). Seasonal variations of erosion rates and processes on hillslopes in western Colorado. *Zeit. Geomorph. Supplementband,* **5,** 215–38.

Schumm, S. A. (1966). The development and evolution of hillslopes. *J. Geol. Educ.,* **14,** 98–104.

Schumm, S. A. (1967). Rates of surficial rock creep on hillslopes in western Colorado. *Science,* **155,** 560–1.

Schumm, S. A. (1970). Structural origin of large Martian channels. *Icarus,* **22,** 371–89.

Schumm, S. A. (1977). *The fluvial system.* New York: John Wiley and Sons. 338 pp.

Schumm, S. A. (1985). Explanation and extrapolation in geomorphology; Seven reasons for geologic uncertainty. *Jap. Geomorph. Union, Trans.,* **6,** 1–18.

Schumm, S. A. (1988). Geomorphic hazards – problems of prediction. *Zeit. Geomorph. Supplementband,* **67,** 17–24.

Schumm, S. A. and Gellis, A. C. (1989). Sediment yield variations as a function of incised-channel evolution. In *Taming the Yellow River: Silt and floods,* ed. L. M. Brush *et al.,* pp. 99–109. Amsterdam: Kluwer Academic Publishers.

Schumm, S. A. and Hadley, R. F. (1957). Arroyos and the semiarid cycle of erosion. *Am. J. Sci.,* **256,** 161–74.

Schumm, S. A. and Khan, H. R. (1971). Experimental study of channel patterns. *Nature,* **233,** 407–9.

Schumm, S. A. and Lichty, R. W. (1963). *Channel widening and floodplain construction along Cimarron River in southwestern Kansas.* US Geological Survey Professional Paper 352-D, pp. 71–88.

Schumm, S. A. and Lichty, R. W. (1965). Time, space and causality in geomorphology. *Am. J. Sci.,* **263,** 110–19.

Schumm, S. A. and Parker, R. S. (1973). Implications of complex response of drainage systems for Quaternary alluvial stratigraphy. *Nature,* **243,** 99–100.

128

Schumm, S. A., Harvey, M. D. and Watson, C. C. (1984). *Incised channels: Morphology, dynamics and control.* Littleton, CO: Water Resources Publication. 200 pp.

Schumm, S. A., Mosley, M. P. and Weaver, W. E. (1987). *Experimental fluvial geomorphology.* New York: Wiley. 413 pp.

Schumm, S. A., Costa, J. E., Toy, T., Knox, J., Warner, R. and Scott, J. (1982). Geomorphic assessment of uranium mill tailings. In *Uranium Mill Tailings Management*, pp. 69–87. Paris: Nuclear Energy Agency.

Scriven, M. (1959). Explanation and prediction in evolutionary theory. *Science*, **130**, 477–82.

Semmelweis, I. P. (1861). The etiology, the concept and the prophylaxis of childbed fever. (Translation by Frank P. Murphy, 1941, *Medical Classics*, **5**, 350–773.)

Sharp, R. P. (1964). Wind-driven sand in Coachella Valley, California. *Geol. Soc. Am. Bull.*, **75**, 785–804.

Shea, J. H. (1982). Twelve fallacies of uniformitarianism. *Geology*, **10**, 455–60.

Shen, H. W. and Komura, S. (1968). Meandering tendencies in straight alluvial channels. *Am. Soc. Civil Eng. Proc., J. Hydraul. Div.*, **94**, HY4, 997–1016.

Shen, H. W. and Schumm, S. A. (1981). *Methods for assessment of stream related hazards to highways and bridges.* Federal Highway Administration Report FHWA/RD, 80/160, pp. 1–86.

Sigma Xi (1986). *A new agenda for sciences preliminary report.* New Haven: Sigma Xi. 118 pp.

Simpson, G. G. (1963). Historical science. In Albritton (1963).

Spieker, E. M. (1965). The nature of geology and its place among the natural sciences. *New York Acad. Sci. Trans.*, **28**, 159–69.

Starkel, L. (1979). Typology of river valleys in the temperate zone during the last 15,000 years. *Acta Univ. Oulu*, **83**, 9–18.

Starkel, L. (1983). The reflection of hydrologic changes in the fluvial environment of the temperate zone during the last 15,000 years. In *Background to palaeohydrology*, ed. K. J. Gregory, pp. 213–55. New York: Wiley.

Stoddart, D. R. (1986). *On geography.* Oxford: Blackwell. 335 pp.

Strahler, A. N. (1954). Historical geology, dynamic geology and recoverability. Unpublished manuscript of paper presented at autumn meeting National Academy of Sciences, Columbia Univ., New York, 14 pp.

Strahler, A. N. (1987). *Science and earth history – The evolution–creation controversy.* Buffalo, New York: Prometheus Books. 552 pp.

Summerfield, M. A. (1982). Macroscale geomorphology. *Area*, **13**, 3–8.

Susser, M. (1973). *Causal thinking in the health sciences: Concepts and strategies of epidemiology.* London: Oxford University Press. 181 pp.

Suzuki, T. and Takahashi, K. (1972). An experiment on rock abrasion by sand impact at moderate speeds. *Chuo Univ. (Tokyo) Faculty Sci. and Eng. Bull.*, **15**, 239–57.

Suzuki, T. and Takahashi, K. (1981). An experimental study of wind abrasion. *J. Geol.*, **89**, 23–36.

Thomas, H. E. (1962). *Effects of drought in the Colorado River Basin.* US Geological Survey Professional Paper 372-F, 50 pp.

Thomas, L. (1974). *The lives of a cell.* New York: Viking Press. 180 pp.

References

Thomas, L. (1979). *The medusa and the snail*. New York: Viking Press. 175 pp.

Thomas, L. (1983). *Late night thoughts on listening to Mahler's Ninth Symphony*. New York: Viking Press. 168 pp.

Thornes, J. B. (1983). Geomorphology, archaeology and recursive ignorance. *Geogr. J.*, **149**, 326–33.

Thornes, J. B. and Brunsden, D. (1977). *Geomorphology and time*. London: Methuen.

Trudgill, S. T. (1976). Rock weathering and climate: Quantitative and experimental aspects. In *Geomorphology and climate*, ed. E. Derbyshire, pp. 259–99. London: John Wiley.

Udvardy, M. D. F. (1981). The riddle of dispersal: Dispersal theories and how they affect vicariance biogeography. In *Vicariance biogeography: a critique*, ed. C. Nelson and D. E. Rosen, pp. 6–29. New York: Columbia University Press.

Van Bemmelen, R. W. (1961). The scientific character of geology. *J. Geol.*, **69**, 453–63.

van der Heijde, P. K. M. (1988). Spatial and temporal scales in groundwater modeling. In *Scales change*, ed. T. Rosswall, R. G. Woodmansee and P. G. Risser, pp. 195–224. New York: John Wiley & Son.

Von Bertalanffy, L. (1952). *Problems of life*. London: Watts and Co. 216 pp.

Waltham, A. C. and Ede, D. B. (1973). The karst of Kuh-E-Parav. Iran. *Cave Research Group, Great Britain, Trans.*, **15**, 27–40.

Waters, M. R. (1985). Late Quaternary alluvial stratigraphy of Whitewater Draw, Arizona: Implications for regional correlation of fluvial deposits in the American Southwest. *Geology*, **13**, 705–8.

Watson, C. C. and Harvey, M. D. (1988). Geotechnical and hydraulic stability numbers for channel rehabilitation: Part I, The approach. In *1988 National Conference Proceedings*, ed. S. R. Abt and J. Gessler, pp. 120–5. American Society of Civil Engineers.

Watson, R. A. (1966). Discussion: Is geology different: A critical discussion of 'The fabric of geology'. *Philosophy of Science*, **33**, 172–85.

Watson, R. A. (1969). Explanation and prediction in geology. *J. Geol.*, **77**, 488–94.

Webb, R. H. (1985). Late Holocene flooding on the Escalante River, South-Central Utah. Unpublished PhD dissertation, University of Arizona, Tucson, 204 pp.

Webb, T. III, Kutzbach, J. and Street-Perrott, F. A. (1985). 200,000 years of global climatic change: Paleoclimate research plan. In *Global change*, ed. T. F. Malone, and J. G. Roederer, pp. 182–219. New York: ICSU Press.

Weinberg, G. M. and Weinberg, D. (1979). *On the design of stable systems*. New York: Wiley. 353 pp.

Westgate, L. G. (1940). Errors in scientific method – glacial geology. *Sci. Mon.*, **51**, 299–309.

Westman, W. E. (1978). Measuring the inertia and resilience of ecosystems. *BioScience*, **28**, 705–10.

Wildman, N. A. (1981). Episodic removal of hydraulic-mining debris, Yuba and Bear River basins, California. Unpublished MS thesis, Colorado State University, Fort Collins, CO, 107 pp.

Williams, G. P. and Wolman, M. G. (1984). *Downstream effects of dams on alluvial rivers*. US Geological Survey Professional Paper 1286, 83 pp.

References

Wolman, M. G and Gerson, R. (1978). Relative scales of time and effectiveness of climate in watershed geomorphology. *Earth Surface Proc.*, **3**, 189–208.

Wolman, M. G. and Miller, J. P. (1960). Magnitude and frequency of forces in geomorphic processes. *J. Geol.*, **68**, 54–74.

Womack, W. R. and Schumm, S. A. (1977). Terraces of Douglas Creek, Northwestern Colorado: An example of episodic erosion. *Geology*, **5**, 72–6.

Woodall, R. (1985). Limited vision: a personal experience of mining geology and scientific mineral exploration. *Aust. J. Earth Sci.*, **32**, 231–7.

Wright, H. E., Jr (1984). Sensitivity and response time of natural systems to climate change in the late Quaternary. *Quaternary Sci. Rev.*, **3**, 91–131.

Yochelson, E. L. (ed.) (1980). *The scientific ideas of G. K. Gilbert*. Geological Society of America Special Paper 183, 148 pp.

Zwart, P. J. (1976). *About time*. New York: American Elsevier. 266 pp.

Index